CISM Courses and Lectures

The series presents lecture notes, monographs, edited works and proceedings in the field of Mechanics, Engineering, Computer Science and Applied Mathematics.
Purpose of the series is to make known in the international scientific and technical community results obtained in some of the activities organized by CISM, the International Centre for Mechanical Sciences.

International Centre for Mechanical Sciences

Courses and Lectures Vol. 543

For further volumes:
www.springer.com/series/76

Armando Di Nardo · Michele Di Natale · Anna Di Mauro

Water Supply Network District Metering Theory and Case Study

Springer

Armando Di Nardo
Michele Di Natale
Anna Di Mauro
Second University of Naples, Italy

ISSN 0254-1971
ISBN 978-3-7091-1492-6   ISBN 978-3-7091-1493-3 (eBook)
DOI 10.1007/978-3-7091-1493-3
Springer Wien Heidelberg New York Dordrecht London

All contributions have been typeset by the authors
Printed in Italy

Printed on acid-free paper

Springer is part of Springer Science+Business Media (www.springer.com)

PREFACE

A water supply network is the key link in the water distribution chain, because it allows for the distribution of the resource for civil, industrial and agricultural uses.

Reported by many sources, both in Italy and other countries (UNESCO, 1993, United Nation, 2000, 2002a and 2002b; National Geographic Italia, 2002; World Water Council, 2003; Ministero dell'Ambiente e della Tutela del Territorio, 2004 and 2005), the waste of water resources in supply networks is considerable.

The deterioration of these systems is due to many causes related to aging technology (average life of European water networks exceeds 30 years) and to negligence and poor organization of the various authorities in charge of ordinary and extraordinary management.

These reasons add to the essential complexity of the urban water distribution systems arising from many technical and scientific problems.

Among technical problems, it is worth mentioning:

- *poor knowledge of the actual plans due to the unplanned growth of the water system along with the urban sprawl, of the 1970's and 1980's;*
- *lack of digitalized plans and TIS (Territorial Information Systems);*
- *scarce availability of up-to-date and reliable databases;*
- *aging of supply networks and hydraulic devices (pipelines, gates, adjustment valves, special works, flow meters, pressure transducers, etc.);*
- *considerable physical and administrative water leakages;*
- *difficulty in defining the two leakage rates ("revenue" and "non-revenue" water);*
- *technological and operational backwardness due to the absence of modern systems of remote-controlling, and to the lack of intervention plans for both ordinary and extraordinary maintenance;*
- *non-compliance, in many countries, with local management "best practice";*
- *complexity of the procedure to carry out hydraulic checks in very large water systems;*

- *poor hydraulic performances (inadequate pressure, reduced resources during drought, poor quality, etc.);*
- *faulty planning of many water systems currently in operation;*

Among the scientific problems, it is worth mentioning:

- *non linear behavior of equations modeling the water distribution systems;*
- *presence of a very high number of variables which makes the application of optimization techniques particularly complicated;*
- *difficulty in defining cost functions due to the lack of data relating to the management of water services;*
- *difficulty in defining optimal criteria for district metering pursuant to management "best practice";*
- *low number of pilot sites;*
- *implementation of case studies on small water networks;*
- *lack of multi-scenario analyses relating to drought;*

Obviously, these issues, especially the technical ones, assume more or less importance depending on the international reality taken into account.

In many countries, as is the case in Italy, there is a clear need for modernization to reach standards of cost-effectiveness: by improving service, updating databases, billing consumptions regularly, creating multithematic TIS, using modern monitoring technologies and criteria and integrated systems to optimize the whole water system.

In order to do so, it is necessary to develop Decision Support Systems, DSS, (Loucks and da Costa, 1991; Loucks, 2000) which help managers by simplifying (thanks to remote controlling, optimization models and operational research) the complexity of water systems in term of hydraulics and customer satisfaction.

Indeed, besides achieving satisfactory hydraulic performance, a water supply company has to fulfill the needs that modern consumer societies demand, and rightly so, i.e., all those elements which make a company more competitive than another and which result from the maximum possible control of the whole system.

In this study two techniques are put forward to improve water supply system management: "Water District Metering (WDM)" and "Water Pressure Management (WPM)" with reference to leakage reduction.

Specifically, after a theoretical introduction of two techniques in Chapter 1 and Chapter 2, the study proposes the use of both techniques in tandem defining two tools for WDM design and WPM setting (Chapter 3). Water district metering and pressure management were already used in some other experiences, often using empirical approaches applied to small water networks.

In this study two original tools are proposed; they are based on heuristic and optimal criteria, arranged in design support methodologies to assist water network operators.

In Chapter 4 the two techniques are tested on a case study of the Monterusciello network in Pozzuoli (Italy) confirming the effectiveness of the approach presented.

The results illustrated in this monograph are part of a research work conducted by CIRIAM (Centro Interdipartimentale di Ricerca in Ingegneria Ambientale) of Seconda Università di Napoli (SUN).

Thanks to Annalisa Lama and Giovanni Francesco Santonastaso for their precious collaboration in the revision of the book.

Armando Di Nardo[1], *Michele Di Natale*[2], *Anna Di Mauro*[3]

---

[1]Assistant professor of Hydraulic Constructions at the Second University of Naples and Ph. D. in Engineering of Civil Networks and Territorial Systems with specializations in Hydraulics and Environment. He has published many papers in italian and international journals on optimization in the management of water resources.

[2]Full professor of Hydraulic Constructions and Dean of the Faculty of Engineering at the Second University of Naples. He has published many papers in italian and international journals on management and planning of water systems.

[3]Enviromental engineer with specialization in Hydraulics.

Contents

1. Water Leakage Management

The greatest problem for a water network - and surely the most important from an economic and social standpoint - is related to water leakages. In many areas of the world more than one third of the resource,which is first extracted, then treated to improve its organoleptic qualities to make it drinkable, and then conveyed to urban water networks, is lost before it reaches end users.

Leakages are a major technical and management problem also for industrialized countries. High leakage levels, sometimes even more than 50% [64] of the supplied volume, are nowadays not sustainable for a water utility, both from an environmental and economic point of view. Some countries have been tackling the leakage problem for many years, promoting basic research, special laws and the implementation of best practices [101]; many others, instead, addressed the problem with enormous delay [62].

The importance of the topic, internationally recognised also by organizations like UNESCO [91], has led many important associations [2, 49] to analyse its technical and scientific aspects in depth, and to promote the dissemination of technologies already available on the market or previously experimented by important water utilities.

Leakages, in a preliminary evaluation, can be divided into real (or physical) losses, that is water actually lost when delivering the service, and apparent (or unauthorized consumption) losses, i.e. water delivered to consumers but not billed [3]. The considerable presence of both kinds, for different reasons, undoubtedly represents a poor physical situation of the water system and the absence of reliable systems of metering and management of the service, with clear effects on the economic, political and social aspect and, in general, on conserving and managing water resources.

For many years a number of international technical and scientific organizations have been concerned with the issue of water leakages [1, 66]. The International Water Association (IWA), in 1996, promoted a specific Task Force to analyze and evaluate the methodologies for the international comparison of water leakages from water distribution systems with the following goals:

- standardization of the terminology for the calculation of water balance and actual and apparent losses;
- the review of possible performance measures for the international comparison of leakages.

The work of the Task Force is summarized in the document "Losses from Water Supply Systems: Standard Terminology and Recommended Performance Measures" [49, 29], which includes, in particular, a standardized methodological scheme to evaluate the water balance relating to water supply systems and represents a valuable reference in order to properly understand and standardize water flows.

This schematization is illustrated in several other documents and sector studies and can be considered as the methodological standard of reference in the international framework.

This IWA document also includes a methodological scheme to understand the water outflows from an adduction and/or distribution system, aimed at properly defining - within the framework of the difference input volumes less billed authorised volumes - the quantities actually lost in adduction and distribution networks, in utility storage tanks, connections and those attributable to unbilled and authorised consumptions, to unbilled and unauthorised consumptions, and inaccurate metering. This methodological scheme, shown in Table 1, can be considered as the international reference for studies and operational management in this field.

<table>
<tr><th>A</th><th>B</th><th>C</th><th>D</th><th>E</th></tr>
<tr><td rowspan="9">System Input Volume (m^3/year)</td><td rowspan="4">Authorised Consumption (m^3/year)</td><td rowspan="2">Billed Authorised Consumption (m^3/year)</td><td>Billed Metered Consumption (including water exported)</td><td>Revenue water (m^3/year)</td></tr>
<tr><td>Billed Unmetered Consumption</td><td></td></tr>
<tr><td rowspan="2">Unbilled Authorised Consumption (m^3/year)</td><td>Unbilled Metered Consumption</td><td></td></tr>
<tr><td>Unbilled Unmetered Consumption</td><td></td></tr>
<tr><td rowspan="5">Water Losses (m^3/year)</td><td rowspan="2">Apparent Losses (m^3/year)</td><td>Unauthorised Consumption</td><td>Non Revenue water (m^3/year)</td></tr>
<tr><td>Metering Inaccuracies</td><td></td></tr>
<tr><td rowspan="3">Real Losses (m^3/year)</td><td>Leakage on Transmission and/or Distribution Mains</td><td></td></tr>
<tr><td>Leakage and Overflows at Utility's Storage Tanks</td><td></td></tr>
<tr><td>Leakage on Service Connections up to point of Customer metering</td><td></td></tr>
</table>

Table 1. Elements of the water balance suggested by IWA

It is worth noting that the item "apparent losses", according to IWA standardization, includes only "unauthorised consumptions" and "metering inaccuracies" whereas technical uses such as washes, cleanings, etc., are not to be considered as actual losses but more properly as technical consumptions connected to the operation and maintenance of networks and included in the item "unbilled authorised consumption".

1.1 Loss control and localization

Effective control of leakage is critical, in some systems, water may have to be rationed in the attempt to distribute the available water more equitably, but frequently such control is given in sufficient priority.

Leakage control practices are very important and are used to maintain a high level of regular monitoring and to repair the leaks which become self evident.

Currently, the most common method of leakage control is *passive leakage control*:only the evident water loss is detected, located and repaired. A leak may be evident because water appears on the road surface or thanks to consumer complaints. This is a passive control method, incompatible with a modern water system management.

That is why recently some different new *active leakage control* methods have been suggested.

All these methods are reviewed hereinafter with reference to the following tasks:

a) evaluation
b) detection
c) reduction
d) monitoring

These tasks, explained below, are not independent of one another; in many cases, only the combined use of various activities produces good results in decreasing the two rates of real and apparent water loss:

a) *evaluation* of water leakage is mainly based on the water balance and on the Minimum Night Flow (MNF) measurement [104], by assigning to the former a leading role in preparing reports on water losses and to the latter a fundamental role in their management. These techniques are also important for a systematic approach to monitor water losses, since they permit a comparison to be made between the different areas of the network and the evaluation of priorities and efficacy of interventions.

b) *detection* techniques of real water losses are mainly of acoustic kind; they are based on the sound waves produced by the water flowing out from a break, whose persistence makes it possible to distinguish it from other noises. Typical listening instruments like geophones and correlators are supported by other radio transmitting acoustic signalers. A common aspect of all acoustic techniques is they allow the water loss only to be located not to be evaluated.

The activities based on the detection of the noise made by water during the leakage can be manual (*manual sounding*) or automatic (*automatic sounding*). Manual sounding is performed by an operator who tracks the network by mobile geophone; this method provides for limited duration of listening and can lead to misinterpretations.

In the last few years, modern techniques of automatic sounding have been developed that allow a fast and cheap automatic prelocation of leakages thanks to fixed geophones. In short, automatic sounding is carried out by installing some noise detectors that, during the night, when the interferences caused by other noises are lower, detect and recognize the "noise" made by the leakage.

The use of automatic sounding techniques allows the leakage localization to be improved through correlation analysis of vibration and noises caused by water leaks.

The noise correlator is based on the principle that a water loss generates an increment in noise, in some cases also considerable, in the water network caused by the interaction between water and pipe material: in the event of leakage, the noise increases in intensity and the leak noise correlator can locate the position of a leakage by comparing two recording between two points of a pipe with a water loss and computing the travel time interval of leakage noise. The leak noise correlator consist of a transducer and a receiver, often used manually by operators that move the tool along the water pipes. In the hypothesis that the pipe material is homogeneous, the leakage noise propagates in the pipe with the same speed and so the measures must have almost the same noise base albeit out of phase in time. The leak noise correlator will be able to recognize the similarity of two signals in both pipe points and detect the time lag that minimizes the difference thereby locating the water loss.

Therefore, the activities used for leakage detection can be divided into two main types [48]; *prelocation* which means identifying the critical areas with possible leaks and the actual *location* which allows the specific critical sub areas of water losses to be identified.

The method applied for leakage prelocation are mainly of two kinds:

- those providing for measurements of the flow through possible valve operations in order to split networks into subsystems and compute the water balance (step test or zooming analysis);
- those based on leakage noise detection (electroacoustic and correlative analyses).

The above-mentioned techniques have advantages and disadvantages and they often need to be combined to get the best economic results and the possibility to monitor the water distribution system at the end of the location campaign and after leakage repairs.

Water utilities, based on their experience, the size and characteristics of the network, use leakage detection techniques according to various methodologies:

- *carpet-correlation*: correlation is performed along the whole network to be examined. Neither prelocation of leakages nor flow size measurements are carried out in the studied area. Carpet -correlation represents the typical technique to locate losses.
- *sounding and correlation*: unlike carpet-correlation, there is a step of prelocating leakages through sound tools along the whole network to be examined. Critical areas are then signalled, where correlation is performed. It is an evolution of carpet-correlation since it focuses on locating leakages where they really are.
- *typical zooming*: it splits the area into districts of about 20.000 inhabitants and is monitored with permanent flow meters. Then critical areas are split into measurement districts and the "step test" is performed. The step test is based on a temporary partitioning of critical "districts" into a variable "n" number of small sub-systems (or "steps") and allows accurate evaluation of the individual areas where leaks are mostly present. In the "steps" (or sub-systems) which present a high rate of leakage, losses are located by sounding and correlation. This methodology is characterized by a heavy network partitioning, though temporary, aimed at identifying the most critical areas.
- *advanced zooming*: it splits the network into districts of about 20.000 inhabitants. Critical areas are identified and radio- transmitting sound sensors are installed uniformly to allow prelocation of losses. Sounding and correlation are performed to locate leakages in those districts showing a higher rate of losses. Advanced zooming is similar to typical zooming until district creation. No further district metering is carried out.

- *new technologies of automatic correlation*: the steps are similar to advanced zooming until leakage prelocation. The sounding and correlation activity is replaced by the installation of special sensors in those areas where losses are expected. These sensors allow a remote automatic correlation to be carried out. This technique is being improved especially as concerns the possible radio connections between remote station and automatic correlation loggers.

Finally there are other techniques to locate leaks, which are based on different principles, like those that studying *pressure waves* [31] or based on *infrared thermography* which, nevertheless, cannot yet be used on a large scale and still present major problems of engineering.

It is worth mentioning another technology: gas detection, which consists in introducing an inert gas into the pipe. The gas is detected through sensors while it comes out at the leakage point. This technique is usually used either for "difficult" losses (in particular: conditions of low pressure and non-metal main pipes) or for losses in new pipes after negative results during pressure tests.

c) the *reduction* of real losses can be achieved by repairing or replacing pipes. Nevertheless, in recent years a different approach has been developed based on controlling real losses by reducing pressure levels in the distribution network, which has a positive effect also on the reduction of breakages. This technique will be explained in detail in section 1.2.

d) systematic *monitoring* allows losses to be kept under control and the effectiveness of repairs to be evaluated. Furthermore, it permits the collection of data relating to significant water quantities, by using remote supports that are fundamental for future projects and water supply system planning.

The short review of the main, typical and non typical, techniques for the prelocation and location of leakages has shown that - since it is necessary to operate along the whole water supply network - the network has to be split into smaller areas (districts) to simplify the evaluation of water balances in order to estimate losses and then monitor them.

Therefore, the latest techniques of research and control of leakages are based on a preventive definition of "district meter areas" of the water network which allows the problem to be tackled more rationally. By splitting the network into districts a reliable preliminary network integrity evaluation can be made(which is hardly possible on the whole system). In this way, it is easier to identify the areas with problems and to understand where the system is more vulnerable and where there are heavy losses.

Water supply network district metering (WDM) thus consists in dividing the water network into districts in order to define different areas that have the same properties and are metered continuously.

This method simplifies the application of the techniques mentioned above (sounding, correlation, step test, zooming) which lead to increasingly heavy partitioning of the whole system. By working on smaller portions that are more easily manageable, also the effectiveness of reduction techniques, such as pressure management, increases significantly.

The following chapters will deal with the main aspect connected to district metering of water distribution networks and the problems relating to its implementation.

1.2 Pressure management

Water pressure management (WPM) is the practice of managing system pressure to a level of service ensuring efficient supply to customers. Pressure is Known to have a great influence on the level of real losses. Pressure decrease can reduce network leakage, limiting the risk of bursts and reducing water losses quantity and providing a constant service to customers. However water pressure reductions are very difficult because all systems cannot tolerate pressure decrease and it is complex to give a weighted average pressure for a specific network area due to variations caused by topography and hydraulic profiles. For this reason, pressure management has to be compatible with hydraulic performance and, specifically, with the "design pressure", defined as the minimum value of pressure to provide a good service for users.

In low pressure systems WPM is a very important type of control [87]. This kind of management includes the implementation of controlled district meter areas, pressure sustaining and level control. For example, in a DMA the control of the pressure even when supply is limited allows the operator to route the excess water to other districts and to guarantee that upstream district pressures are not reduced in an uncontrolled manner.

However, WPM in high pressure systems is the best way to reduce leakage, pressure has a considerable effect on water leakage due to a decrease in flow rates from leaks. High water pressure contributes to burst water mains, cause leaks and water waste. Hence in this kind of systems lowering the pressure can attain benefits, such as the reduction in:

- water leakage with lower running costs and higher water savings;
- water bills for customers;

- pipe breaks with consequent reduction in the number of service interruptions and repair costs;

In order to understand the effect of WPM, real losses can be described with a value L given by the following function of the local pressure P [43, 51]:

$$L = c\,P^{\gamma} \tag{1}$$

in which the values of the coefficients c and γ depend on site-specific conditions.

The Literature in this respect shows that the value of γ coefficients varies considerably depending on specific field or laboratory conditions [47, 52, 84, 20]. Indeed the relationship page 12 is sometimes applied to a single pipe, sometimes to a district of the water network, with consequent unclear definition of the above parameters.

In the former case, experimental studies have shown that it can be considerably larger than 0.5 and typically varies between 0.5 and 2.79 [30] or between 0.42 and 2.30 [38], these differences being mainly related to burst type (round hole, longitudinal crack, circumferential crack, corrosion cluster, etc.) and to the pipe material (plastic, steel, concrete, etc.).

In the latter case, a similar pressure-leakage relationship is employed [87], based on field data from the UK, Japan, Australia, Brazil, Canada, Malaysia, New Zealand and the U.S.A:

$$\frac{L_1}{L_2} = \left(\frac{P_1}{P_2}\right)^{N1} \tag{2}$$

The $N1$ exponent was found to vary between 0.5 and 1.5, but it may occasionally reach values as high as 2.5. These results can also be used for the prediction of $N1$ in the absence of specific test data [86]. From relations page 12 and page 12 it is evident that WPM, avoiding un-needed pressure excesses, may reduce pipe leakage during the day, with the twofold reduction of water leakage [32, 43, 57] and burst frequency.

Estimation in equations page 12 and page 12 of c, γ and $N1$ values is therefore required for any leakage control and management policy. In the chapters below a method to determine the two parameters, based on a genetic algorithm, is proposed, which relies on the simultaneous detection of demand and the leakage pattern in a water distribution network.

Therefore, it is clear from page 12 that the placement of valves to reduce pressure, in properly chosen locations, can lead to a distribution of water flows at the nodes such as to reduce leakages from pipes. This pressure decrease, compared with the design pressure, inevitably reduces the network's hydraulic efficiency which has to be limited by numerical or empirical optimization of the placemenent and adjustment of the degree of valve opening.

Furthermore, if pressure adjustment is made on distribution networks that have already been partitioning, pressure could already be reduced due to the worsening resulting from the partitioned of the water system.

Nevertheless, the possibility of adjusting the pressure of each district of the partitioned network [4, 19] may lead to an optimal WPM, depending on the height and kind of the water network of the built-up area involved, by keeping satisfactory rates of hydraulic performance. Pressure adjustment is made according to the different operational conditions of the water network (peak, average, etc). The best efficiency is achieved during the night when water flows are considerable, and often excessive, whereas the inconveniences resulting from pressure reduction are barely perceptible due to lower water consumption.

2. Water District Metering policy and practice

Water supply network District Metering (WDM), a method developed in England [104] and already implemented in many countries, as illustrated below, aims at:

a) simplifying network water balance calculation by monitoring night flows in each district in order to detect the presence of unreported bursts and to enable leakage detection and location (using acoustic methods, step test, etc.);
b) carrying out pressure management in order to reduce hydraulic head and water leakage;
c) improving water system management with continuous monitoring of district hydraulic data in order to prevent water shortage and to plan better maintenance operations.

As recommended in the literature [101], the water balance technique aimed at leakage monitoring requires the installation of flow meters at strategic points throughout the WDS.

In this way water balance monitoring is simpler, can give precious information about district water losses, provides the basis for network operation and helps prioritize districts for leak detection and location.

District monitoring is carried out by calculating the difference between inflow and outflow of each district, measured by flow meters located at different points in the network.

Flow meters may sometimes contain a pressure valve in order to achieve water pressure management. In these situations the division of the network in hydraulically independent districts [4] allows more effective pressure regulation since a different head level can be obtained in every single network subsystem by inserting boundary (or isolation or gate) valves.

Water district metering can be carried out in different ways and with various partitioning levels [101]: in particular, WDM can be permanent or temporary, with reference to time duration, and can have different sizes based on its specific objectives (network management, water balance, pressure regulation, etc.).

The best results in leakage monitoring can be achieved by defining a small permanent district, called a District Meter Area (DMA), provided that a thorough water system analysis is carried out. Depending on the characteristics of the network, DMAs may be supplied by single or multiple feeds, may flow into adjacent DMAs or be self-contained. The fewer the flow meters the easier and cheaper the calculation of DMA water balance; this can be obtained by using boundary valves to reduce the number of pipe connections according to hydraulic constraints and performance.

In particular, with reference to DMA, it is possible to fix the lower limit of permanent portioning of the water supply network by indicating that area of the distribution network created by permanently closing valves where the quantity of water inflow and outflow is constantly measured as shown in Figure 1.

If district metering is temporary it is called *Waste metering*, illustrated in Figure 2; in this case the water network can be more easily divided into defined areas that have the same properties and are generally smaller than DMAs.

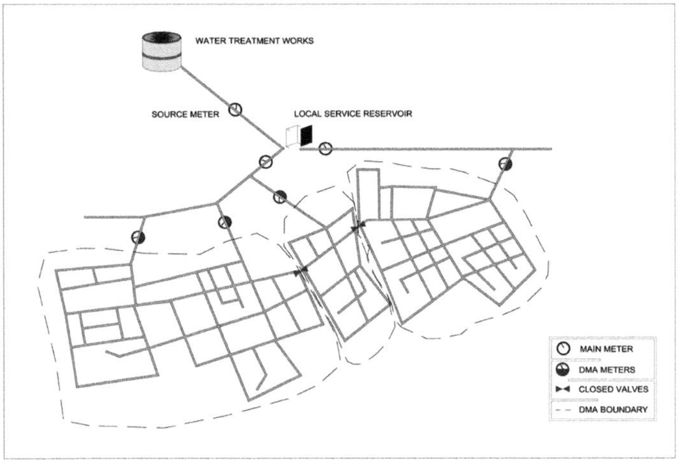

Figure 1. Permanent district metering stuctures (District Water Metering)

The leakage is determined by measuring and recording the low rate of flow that occurs at night-time, called Minimum Night Flow (MNF) [104]. Waste metering is characterized by intense district metering, thought temporary, aimed at identifying the most critical areas. As MNF is recorded in each district at regular intervals, when a strange value is recorded the inspectors are directed to a specific district of the network.

Often district and waste metering are used together: when an unexpected value is recorded in a DMA, the waste meters are used to subdivide

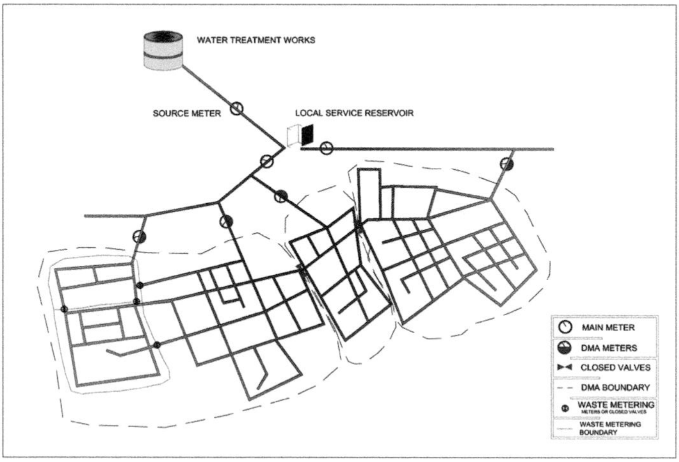

Figure 2. Temporary district metering stuctures (Waste Metering)

the district into more controllable units. Thus the inspectors are guided to the areas containing the largest number of losses in order to locate and identify leakage positions. In this way WDM improves significantly emproves evaluation and detection presented in the previous section.

2.1 Physical and Virtual District Metering

If WDM is carried out in the water supply network design phase, different districts may be planned in order to improve system management, pressure regulation and water leakage detection without altering the network performance. When water systems are already in operation it is far more difficult to set the number and size of districts and, in many cases, WDM may even seriuosly affect the hydraulic performance of water networks both in terms of water quality (creating more dead-end pipes and flow reversals) and node heads [104].

Clearly that any district metering action on the network through flow-adjustment valves has a negative impact on the performance of water distribution system and on water quality. Operating by manoeuvering and closing valves in a difficult and complicated context like urban distribution networks, which have often developed without any preventive planning, causes many technical and economic problems.

DMA planning is simpler for branched networks, while it is much more complicated with looped networks that are typical, for example, of Italian [6, 11, 15, 16, 17, 33, 25] and North American [84] residential areas.

Permanent WDM changes the original topological layout of water systems and reduces network water pressure, especially during peak demands. Though positive in terms of water loss decrease during night flow, it may lead to insufficient supply to customers during peak demand hours and, in this sense, diminish the level of service.

WDM changes hydraulic behaviour since, in principle, it is in conflict with traditional design criteria [79] of looped water networks which allow water systems to be more reliable under mechanical and hydraulic failure conditions [56]. The main design criterion is pipe redundancy, this markedly improves network reliability which can be significantly reduced by pipe closure required to obtain permanent WDM (that can also be generically described as "a permanent failure").

Water District Metering may cause an alteration of general and partial hydraulic performance of a water system. Thus network districts should be designed by experts with the aid of simulation software, especially for extensive water supply system analysis comparing different possible solutions. Many technical and economic constraints have to be taken into account (topology, presence of boundary valves, available budget, etc.) in order to define permanent District Meter Areas that do not diminish the hydraulic performance of the water distribution network.

Furthermore DMAs have to meet the needs of planners (areas with different pressure levels, compliance with levels of service required, etc.) which are difficult to describe in mathematical terms. Therefore optimization methods based on the definition of a precise multi-objective function cannot be easily applied.

Actually WDM depends essentially on the skills of the planner who must identify the best possible solution among the diverse district metering options available. In particular, it is possible to define a "physical district metering" and a "virtual district metering" by proposing also a different terminology classification [15, 16, 17].

In the first case, the adjective "physical" refers to the real action that district metering - through the operation of isolation valves and measurement tools - performs upon the urban water distribution system, by altering its operation in terms of hydraulic performance and water quality.

In the second case, the adjective "virtual" means that the system's district metering only occurs in terms of measuring the hydraulic characteristics (flow, pressure, etc.) of a partitioned area without activating any interception device, hence not harming its operation. Thus virtual "district metering" is carried out by partioning the network through the installation of devices metering flows and bidirectional volumes along the pipes selected to limit the district.

All other possible techniques of district metering - carried out by using more inflow and outflow meters and boundary valves in other interconnection points - will represent "hybrid" techniques of district metering, presenting various problems attributable both to physical and virtual district metering techniques.

Physical district metering gives more opportunities than its virtual counterpart, because it allows the monitoring, managing and optimizing of the operation of the system and the reduction of leakages by WPM thanks to district isolation valves.

These operations need the mathematical model used in simulations to reflect the physical condition of the network as much as possible. Therefore, it is necessary to know a number of data such as: characteristics of pipes, their connections, location of gates, pumps, tanks and hydrants, ground heights, location of users, etc, which are often difficult to collect.

Furthermore, in order to be highly effective, adjustment has to be performed in various operating situations (conditions of minimum, maximum and average consumption), which causes sharp increase in calculation times.

In order to carry out the physical district metering of the network, it is necessary to comply with the following main constraints that, however, vary according to the goal of district metering:

- guarantee the reliability of water supply to individual districts;
- to guarantee water quality;
- to comply with the limits allowed to the changes in the operation of hydraulic plants, also with reference to the availability of water resources and energy optimitazion.

"Virtual" district metering can be carried out by using flow meters, to be installed at each point of interconnection with other districts instead of sectorization valves. Thus, the closing of network district boundaries can be considered "virtual", and it has the advantage of keeping the operational condition of the water distribution network unchanged.

In virtual district metering, the choice of the points where meter devices have to be installed can be made following the indications of an expert, and it is not necessary to carry out simulation or adjustment operations.

This kind of district metering does not give the possibility - at a first stage - to manage the system. However, the network can be monitored, water balances made and, their integrity evaluated through a technique like *zooming*, the prelocation of water losses can be carried out. "Virtual district metering" also raises a number of problems related to the installation of modern metering devices in all points where the district has to be intercepted:

- synchrony of all available sizes;
- bidirectional metering of flows;
- possibility of making measurements available on a single hardware platform;
- availability of software that can calculate water balances in a single virtual district as well as in the whole network;
- possibility of making estimates of unavailable measurements;
- availability of modern and reliable meter devices.

2.2 District Metering design criteria

As explained in the previous sections, district metering (both physical and virtual) of water systems already in operation is complicated and can be carried out according to different approaches which have to take the following aspects into account:

1) *Defining WDM goals*

   Some goals of district metering can be classified and they are, as explained earlier, the monitoring and control of the network, as well as the evaluation or identification of water losses. Their exact definition by operators is fundamental to choose the type (physical or virtual), the time duration (permanent or temporary) and the size of districts.

   The main goals that can be achieved through district metering at different levels of the network are the following:

a) leakage evaluation and detection:

   - evaluation of network integrity by calculating the water balance in each district;

   - prelocation of losses through zooming and/or step-test techniques;

b) leakage reduction:
   - reduction water losses through district WPM;

c) water system monitoring:

- by recording water flow and pressure of each districts, system knowledge significantly improves . Abnormal conditions can thus be highlighted, maintenance interventions can be better planned, etc..

2) *Definition of district number and size*

When the goals have been defined and WDM type and time duration have been chosen, it is necessary to define the district number and size. Specifically, starting from the whole water supply network, districts can be designed with decreasing size in terms of population, network length, number of users connections, etc. depending on WDM goals.

The network topology plays a fundamental role in defining district metering number and dimension; for example, WDM of tree networks is easier than WDM of one looped network.

3) *Selecting optimal network dissection and/or metering points*

The selection of boundary valves or flow meter localization depends on many factors: first of all, the kind of district metering, whether physical or virtual, which determines real disconnections on the water distribution network or only metering disconnections, but it also depends on the topology and on the district metering level.

Another important factor in the selection of disconnection points is the kind of devices that have to be installed along the network, which will obviously depend on the selected goals.

Obviously, if the goal is to design a remote monitoring and remote control system for the water distribution network [14, 21], some kinds of meters will be essential (for example, bidirectional meters) and boundary valves (variable closing), which may result in a different morphology of the districts in comparison with the case, for example, of district metering aiming to merely evaluate network integrity - where unidirectional and fixed isolation valves are sufficient.

Once the goals and the kind of district metering have been selected, the levels have been defined and meters have been chosen, the placement will mainly follow two criteria:

- an empirical approach;
- an approach based on numerical optimization techniques [12, 44, 95].

The selection of one of the two criteria will depend on various and multiple technical and economics consideration, on the importance and size of the network, the importance of the problem to be tackled, etc.

It should be pointed out that the latter approach (numerical optimization) does not always include the former (empirical approach), because all the aspects of the problem cannot be formalized in a single system of mathematical constraints to define a process of numerical optimization.

With regards to the empirical approach, it is possible to define some simple criteria for dissection, based on experiences and knowledge of hydraulic engineering, the scientific literature, simulations carried out in similar cases and mere of economic considerations.

In particular, partitioning should be performed where:

- the gradient of the quantity to be measured or controlled is higher;
- it is possible to bound areas with homogeneous pressure;
- it is possible to bound areas where the vulnerability analysis has shown higher risks of breaks;
- there are already isolation and/or measurement devices;
- there is compliance with the maximum number of inhabitants suggested according to the level and kind of district metering (DMA, WM, etc.);
- it is possible to preserve the hydraulic reliability of the partitioned water distribution network;
- it is possible to bound areas with homogeneous materials for pipes;
- it does not worsen the quality of the water delivered.

It is worth stressing that an empirical kind of approach is totally inappropriate due to the complex and non linear nature of the hydraulic equations supporting the system, to any preliminary remark about the global hydraulic operation of the partitioned network after the placement of physical isolation devices.

In other words, the empirical placement of dissection and adjustment devices does not allow reliable prediction of how the various portions of the network, identified as districts or sectors, behave in terms ofwater pressure, flow and quality, especially when networks are heavily meshed. Notwithstanding the above, this approach can be useful during the first step of the zooming process and in all cases of temporary district metering of the network to prelocate water losses.

Then the definition of the appropriate size for a DMA thus depends on a number of system characteristics: demographic condition, required economic level of leakage, leakage control technique and hydraulic condition. Several suggestions about DMA size can be found in the technical literature: in Water Authorities Association and Water Research Centre (1985) a DMA has to include 1,000 - 3,000 properties, while in Butler (2000) a permanent district has to contain 2,500 - 12,500 inhabitants with 5-30km of water network, UK Water Industry Research recommends a number of properties between 1,000 (small DMA) and 3,000 (medium DMA) and up to 5,000 (large DMA). Being based on empirical considerations, and sometimes on too small a number of test cases, these guidelines cannot be easily extended to large water supply systems. Besides the size of a districtits morphology is also very important.

Depending on the morphology of water network, the position of the district in relation to the network and the number of links belonging to the district may affect the efficient supply to customers.

2.3 Experiences in the world

In some countries WDM is used to improve water network management and, specifically, to control water leakage. In this section we report some recent international experiences about the use of water district metering; all these case studies show an empirical approach to define water districts.

a) The case of Canada. Use of Flow Modulated Pressure Management in York Region, Ontario, Canada

The Regional Municipality of York, just north of the City of Toronto in Ontario (Canada) understood the benefits of demand side management initiatives and decided to incorporate Demand Side Management into its long-term water supply strategy. The strategy selected by the Region of York in order to address and reduce water leakage within their water distribution systems was to implement District Metered Areas (DMAs) and Flow Modulated Pressure Management Areas (FMAs).

This methodology is relatively new in North America for active leakage control but is also promoted as the most effective method available [47]. To enable efficient control of recoverable leakage, the distribution systems that make up York Region were divided into 65 temporary DMAs. These DMAs are being used both to identify and reduce recoverable leakage in the short term and then to monitor and control leakage in an ongoing manner.

In addition to the 65 temporary DMAs, 10 dynamically controlled permanent pressure management schemes were introduced to allow unnecessarily high pressures to be efficiently controlled, to reduce background leakage volumes and new leak frequency, while continuing to meet hydraulic and fire fighting requirements.

The DMA program, comprising the 65 areas and targeting unreported recoverable leakage, has yielded results of 7.65 MLD of sustained water loss savings - exceeding the 5.21 MLD target by 47%.

Similarly, the FMA program has also exceeded targets. Two of the FMA schemes operate in the same isolated area within Richmond Hill and are referred to as RH 6-1 & RH 6-2 - Richvale Area. The pressure management area comprise 45 km of watermain with a total of 4250 mostly residential connections, with some small commercial and school sites. The area has an average night-time zone pressure of 56 m. The projected pressure scheme highlighted an average reduction of 15 m across the isolated area with a resulting estimated savings of 8% of average annual daytime demand.

The results of this area support the expected savings calculations. To date they are implementing an average reduction of 10 m with average savings of 6.5% of average annual day-time demands. The full 15 m reduction will be set for the final year of the programme to enhance the water savings.

To date they have achieved 164% of the total project goal of 0.73 MLD with respect to the flow modulated pressure schemes with only 60% of the sites in full operation.

b) The case of Korea. A Case Study of Leakage Management in the City of Busan, Korea

In some parts of South Korea, water loss in the supply network reaches almost 30% the main causes of which include breaks of aged pipes. Korea is now in timely need of fixing this leakage problem, and is considering operational pressure control coupled together with creating a block system [80] as an option.

The block system approach basically converts a large network into a group of small and simple sub-networks which can be more or less independent of each other. It was moreover proved that such a block system can also cope with leakage management since it is easier to search for the causes of water losses and fix in small simple networks than in large complicated networks. The block system is known as an ideal water network system for leakage management. Its positive attributes include the followings: water losses are efficiently controlled by checking flow quantity over a small area; physical losses are reduced because water pressure fluctuation becomes less sharp.

In Korea two strategies were applied: checking metering devices to reduce commercial losses and a series of tests to reduce physical losses. In this application, an area was selected to test the approach as a means to control and manage leakage prior to extending it to the entire City of Busan.

For this test, billed unmetered consumption and unbilled unmetered consumption were ignored because these values were very small compared with the others. Also, unauthorized consumption was assumed to be zero. From the test results, it was found that commercial losses decreased from 122.7m^3/day to 116.4 and physical losses considerably decreased up to 40.5%.

Total reduction of water losses in the project area was 216.6m^3/day (i.e. 33.8% of initial water losses) and the City was able to save about US $ 395,000 per year only in the 2nd block. Encouraged by this result, the City is now positively considering applying the approach to the entire city networks for leakage management.

c) Reducing leakage in Jakarta, Indonesia

Jakarta, the capital city of Indonesia, loses around half of its water supply from leaks in the pipes. Low operating pressures, non-metallic pipes and high background noise make the application of acoustic instrument impossible. A step-by-step approach, based on directly quantifying the leakage, was developed which proved very successfully.

The importance of pressure control to maintain a low leakage level in the network was highlighted. A step-by-step approach based on the direct measurement of the leak was developed. The approch involves in dividing the network into a number of Permanent Areas supplied by a few key mains on which flow meters are installed [74].

These areas were much larger than the more traditional District Meter Areas (DMA) as they cover around 100 km of network. But in the same way as a DMA, they serve to quantify regularly the leakage level and to identify the presence of new leaks. In those Permanent Areas where the specific leakage was high, the network was divided into Temporary Districts, each supplied by a single pipe in which a temporary insertion flow meter is installed. The districts were temporary as the creation of the boundary can cause localized pressure problems. It is thus possible to pinpoint the part of the Permanent Area with most leakage where a night-time step test can be undertaken to identify leaky pipes.

One of the difficulties in the Jakarta network was the inaccuracy of the mains records. When combined with the very low operating pressures of 10 meters or less, this indeed creat a very difficult permanent control system very difficult indeed. In Jakarta this was overcome by building calibrated mathematical models to identify the anomalies, which were then investigated fully in the field by undertaking hydraulic tests and selective excavations.

The Jakarta project has also showed the importance of controlling pressure in networks where there is very little pressure in the first place. The importance of lowering pressure to reduce the amount of water lost in a burst has been understood for a long time. What is less well understood is that high leakage will cause low pressures. So when the leaks are repaired, the pressure will rise, increasing the risk of new leaks forming.

The solution is the installation of a pressure reducing valve (PRV) which will compensate automatically for the increase in pressure, thus ensuring that the lower leakage level can be maintained in the future. With the application of a PRV controller, it is possible to further lower the night pressure, with consequential reduction of the leakage. So successful has the Jakarta pilot project been, that it is currently being extended to cover all of the 3000 km of network managed by Palyja. Not only will this yield a significant reduction in the leakage level, but more importantly perhaps, enable the extremities of the network to receive a continuous supply of water.

d) Experiences in DMA redesign at the Water Board of Lemesos, Cyprus

The Water Board of Lemesos operates a well-organised supply and distribution system with permanent pressure zones and district metered areas, thus providing a solid foundation on which an effective leakage control policy has been developed [19]. The Water Board valued apparent losses and real Losses. Apparent losses are valued at retail billing rates whereas the real losses are valued at the variable cost of water production and distribution. Therefore the Water Board considers that it is as important to reduce of apparent losses as real losses.

The Water Board operates seven pressure zones each fed by gravity from a dedicated storage reservoir. Since 86% of the flow is in pressure zones, it was considered important to first carefully examine the size of the DMAs in pressure zones in an effort to further reduce the real losses from the system and able to provide better and more effective active leakage control.

Each DMA is provided with an inlet chamber which houses a strainer for meter protection, the district water meter, a pressure reducing valve and a pressure sensor.

Continuous flow monitoring and management of pressure began immediately upon completion of the redesign works in each DMA. This has long been recognized by the Water Board and the ultimate goal is for all DMAs to be equipped with PRVs to reduce pressure where possible and to control and stabilize pressure in DMAs where pressure reduction is not practicable. Pressure reduction was applied to all the redesigned DMAs in the pressure zone, taking into consideration the minimum standard of service as outlined in the DMA management section above and the MNF recorded.

The DMA redesign and the application of pressure reduction produced favourable results with both background leakage and locatable losses being reduced by approximately 38%. Furthermore the frequency of reported leaks was reduced by approximately 41%. The overall pressure reduction for the 15 DMAs under consideration was of the order of 32%.

The target of the Water Board of Lemesos is to reduce the NRW to about 8% of the system input volume, which is considered to be the economic level of leakage. The Water Board demand forecasts indicate an increase of approximately 30% by the year 2020. Leakage reduction will go some way towards offsetting this increase in demand as well as provide considerable cost savings.

Finally, it is worth mentioning that also the European Union has already focused on the issue of waste reduction and optimization of water supply network management in recent years, and other important projects involving several countries have been carried out. In particular, it is worth mentioning CARE-W, HYDROPLAN-EU and TILDE projects.

The **CARE-W** project (Computer Aided REhabilitation of Water networks) involves 24 participants among which 11 institutions such as universities and research centres, located in seven European countries and aims to supply public utility companies with a full set of tools designed to improve the management of water supply networks through the creation of Decision-making Support Systems (DSS).

The importance of these tools based on a preventive approach is that they "suggest" to managers the best time to intervene in maintenance and with the most effective technology before an inefficiency occurs. The project aims to tackle the issue of reliability of water networks and urban drainage at European level, in significant infrastructural and economic terms (http://care-w.unife.it/intro.html).

The **HYDROPLAN-EU** project, coordinated by Aquaplus nv (a Belgian company controlled by Aquafin nv), involves five pilot sites:
Belgium (Louvain), Greece (Athens and Thessaloniki), Ireland (Meath) and

Italy (Imperia) - Given that many international statistics show that 20% of technological networks account for 80% of malfunctions - the project aims to steer the management of technological networks towards the subset which causes the highest amount of damages, in order to optimize investments and intervention costs on the system.

The project uses GIS systems intensively as an analysis tool and it is based on the construction of a matrix of damage and risks (structural, hydraulic and environmental) to identify the portions of networks which are most likely to show inefficiencies and where, therefore, it is necessary to carry out a rehabilitative intervention (http://www.hydroplan-eu.com/).

Finally, the **TILDE** project (Tool for Integrated Leakage Detection) involves leading companies and research centres in Europe. It provides for the definition of Decision-making Support Tools (DST) for a more efficient management of water networks with special attention to the problem of leakage reduction.

Starting from the awareness that there is a very large margin to recover water leakages, TILDE studies the extent to which the economic and environmental value of the water resource - related to individual local reality - affects the cost-benefit analysis of a recovery campaign for water leakages.

Therefore, TILDE seeks to connect the social and economic aspects to the sustainable management of water leakages, both in economic and environmental terms. The DST of TILDE, developed according to a concept of industrialization of procedures, in contrast with the traditional non-systematic practices still currently used by most managers, aims to provide solutions based on the specific needs of operators and local conditions [40].

3. Design Support Methodology

This study proposes a Design Support Methodology (DSM) to define Water District Metering compatibly with water network performance and to determine the optimal setting pressure values to be assigned to valves during the various operational conditions in order to carry out pressure management. In this chapter two method are illustrated: the first, based on techniques borrowed from graph theory [7, 8], that allows several different network configurations to be analysed and compared by means of some performance indices; the second, based on genetic algorithms, that allows description of a non-linear optimization problem in order to define valve optimal settings.

3.1 Water District Metering Design

Water network configuration problems, in particular network topology and connectivity, are classified as *Layout Problems* in the literature [83, 36, 68, 34].

There are not many papers on layout optimization in the literature, and they focus mainly on water system robustness and reliability [96], and not specifically on WDM.

Several scientific works on layout problems use graph theory to analyze water network topology (node plano-altimetric position), connectivity (the probability that a given demand node is connected to a source), reachability (the probability that all nodes are connected to a source), in order to investigate network vulnerability with respect to expected operations (valve regulations, pipe substitutions, etc.). The first applications of this theory to water supply networks, already applied to other network types (electricity, IT, etc.), date back to many years ago [46].

In [41], graph theory is combined with integer goal programming in order to optimize water supply network redundancy and thus improve system reliability by maximizing network regularity. This goal was achieved by minimizing the sum of the deviations at each node in terms of the number of links incident upon it.

In [67] the concept of a water network *backup subsystem* is introduced, a subset of links of the full system where a prescribed level of service is maintained when a failure occurs. The backup subsystem is obtained using algorithms based on graph theory and topologic considerations; while in [68] a frequency connectivity digraph matrix, that maintains Kirchoff's Laws 1 and 2, is used.

In [77] graph theory is used to partition the water network into "tree" and "co-tree" sets to obtain simpler network layouts and to better describe the optimisation problem of minimising the heads by setting hydraulic regulation valves.

More recently in [26] a similar approach to water supply system control and management has been proposed, based on the division of the network graph into two subsystems: a main graph (called "core") consisting of looped pipes and a secondary graph (called "forest") consisting of tree pipes; both graphs are linked by connection elements called "bridges". Using these methods network modelling may be simplified and different views of the hydraulic system may be obtained.

More specifically in a simplified scheme of a water supply network, made of core, forest and bridge, or tree and co-tree sets, it is possible to better understand the interactions between the different parts of the system such as their connection to network sources (reservoir, tank, pump, etc.) and, in some cases, to better define optimization problems.

As to the specific problem of optimal district partitioning of a water distribution system with several reservoirs, [90] suggest an algorithm derived from graph theory to identify independent supply sectors (or districts) of the network layout.

In this study a Design Support Method (DSM) is put forward to partition a water supply system into DMA, by resorting to some graph theory principles that allow analysis of the minimum energy paths computed from each reservoir to each node of the water network.

The DSM was assessed by using performance indices (statistical and hydraulic) to compare the level of service of several district layouts. Specifically, the degeneration of a looped network due to Water District Metering and, as a consequence, the diminished capacity of the water system to react and to overcome stress conditions, was measured with resilience indices [89], based on an energy approach and used in order to compare different WDMs with Original Network Layout (ONL).

The number of possible WDMs of a large water supply network may be huge; nevertheless, many of them are not compatible with the constraints and can undermine the level of service for users. Each district configuration requires that a certain number of boundary valves and several flow meters be inserted into the network to analyse the water balance.

Therefore, for each district, there is a cost to purchase, install and maintain the equipment which, obviously, has to be reduced as much as possible in compliance with the constraints and fixed goals. Furthermore, with an equal number of districts, it is better to have the lowest possible number of flow meters in order to make it easier to evaluate water balances, which need the synchronous comparison of all the measurements. Clearly, the best condition would be to have one single inflow meter for each district so as to make it simple for the operator to calculate the water balance.

The above comments go to show that, among the many possibilities to partition the network into districts, only some of them can fulfil the goals, take into account technical and economic constraints, and be compatible with the demands of users.

Nevertheless, in the case of very large water supply networks, although the number of allowable configurations is limited, it is difficult to find compatible WDMs by following only mere empirical remarks from the literature, even if used together with hydraulic simulation software. Indeed, in the case of complex networks, the set of possible configurations is huge and cannot be investigated with empirical approaches such as "trial and error".

On the contrary, the methodology proposed offers a procedure which is able to rapidly identify a configuration of permanent districts of a water supply network which fulfils goals and constraints. In particular, it supports designers when defining DMAs, by giving them the possibility to compare several possible options among those compatible with the performance of the system.

The approach used was to define districts by identifying those pipes selected to insert boundary valves and flow meters, by searching minimum energy paths. This makes it possible to rapidly find WDMs which are compatible with the levels of service and the reliability of the Original Network Layout (ONL).

Specifically, the method consists in identifying "less important" or "more redundant" pipes in terms of water system reliability that can be closed to create DMAs. The aim of this approach is to reduce the number of pipes among which the choice is to be made as to where to insert flow meters and boundary valves in order to obtain Water District Metering compatible with a good hydraulic performance measured by specific indices.

This method by means of some graph theory principles, allows rapid identification of the more important network pipes that cannot be closed but may be used to insert flow meters for water balance.

The approach proposed is arranged in a Design Support Method on which operators can rely during the whole design process with the following steps (as illustrated in DSM flow chart of Figure 3):

a) simulate WDS to carry out pipe flows q_j and node heads h_i;
b) define adjacency A, incidence I, weight W matrices;

c) compute all Shortest Paths $\{p\}_{min}^{s}$ and all path frequencies f_j ;
d) define Main Network Layout (MNL);
e) draw Main Graph G_M;
f) choose DMA number N_k and size;
g) insert N_{fm} flow meter;
h) close $\{\nu_l\}$ DMA links with boundary valves;
i) simulate WDS again;
l) compute Performance Indices (PI);
m) determine PI and N_{fm} :if PI is not satisfactory return to step f) if N_{fm} is not satisfactory go to step n);
n) delete one flow meter and return to step h);
o) choose WDM.

Specifically, starting from network model INPUT (with n nodes, identified by labels α with $i = 1, \ldots, n$, and m pipes, identified by labels β, with $j = 1, \ldots, m$, and node water demand distribution Q_l with $1 = 1, \ldots, n$, source heads H_s, identified by labels σ_s with $s = 1, \ldots, r$ reservoir, pipes lengths L_j and node elevations e_i, pipe flow q_j, node heads h_i and head loss ΔH_j for each pipe can be calculated by a Demand Driven approach [88] with a hydraulic simulation (step a) . The next step b is the definition of square adjacency matrix A of order n and incidence matrix I of order $n \times m$[9]:

$$A_{xy} = \begin{cases} 1 & \text{if node } \alpha_x \text{ is linked to node } \alpha_y \text{ (for } x \neq y) \\ 0 & \text{else} \end{cases}$$

$$I_{xy} = \begin{cases} 1 & \text{if node } \alpha_x \text{ is the start node of pipe } \beta_y \\ 0 & \text{if node } \alpha_x \text{ is the end node of pipe } \beta_y \end{cases}$$

and of sparse weight square matrix W of order n:

$$W_{xy} = \begin{cases} w_{xy} & \text{if } A_{xy} = 1 \text{ and } w_{xy} > 0 \\ 0 & \text{else} \end{cases} \tag{3}$$

with the following weight function computed between two generic network nodes α_x and α_y:

$$w_j = w_{xy} = q_j \, \Delta H_{xy} = q_j \left(H_{\alpha_x} - H_{\alpha_y} \right)$$

Then, compatibly with network layout $\forall\ s, i$ is possible to identify all node paths $\{\mathrm{p}_z\}^{s\,\alpha_i} = \{\sigma_s, \ldots \alpha_x, \ldots \alpha_i\}^{\,s\,\alpha_i}$ belonging to path set

$\{\mathrm{p}\}^{s\,\alpha_i} = \{\mathrm{p}_1{}^{s\,\alpha_i}, \ldots \mathrm{p}_z{}^{s\,\alpha_i}, \ldots \mathrm{p}_{np}{}^{s\,\alpha_i}\}$ with z=1..$\mathrm{n}_{s\alpha_i}$, that may be crossed by an infinitesimal flow dq from source s to the i-th node in the worst

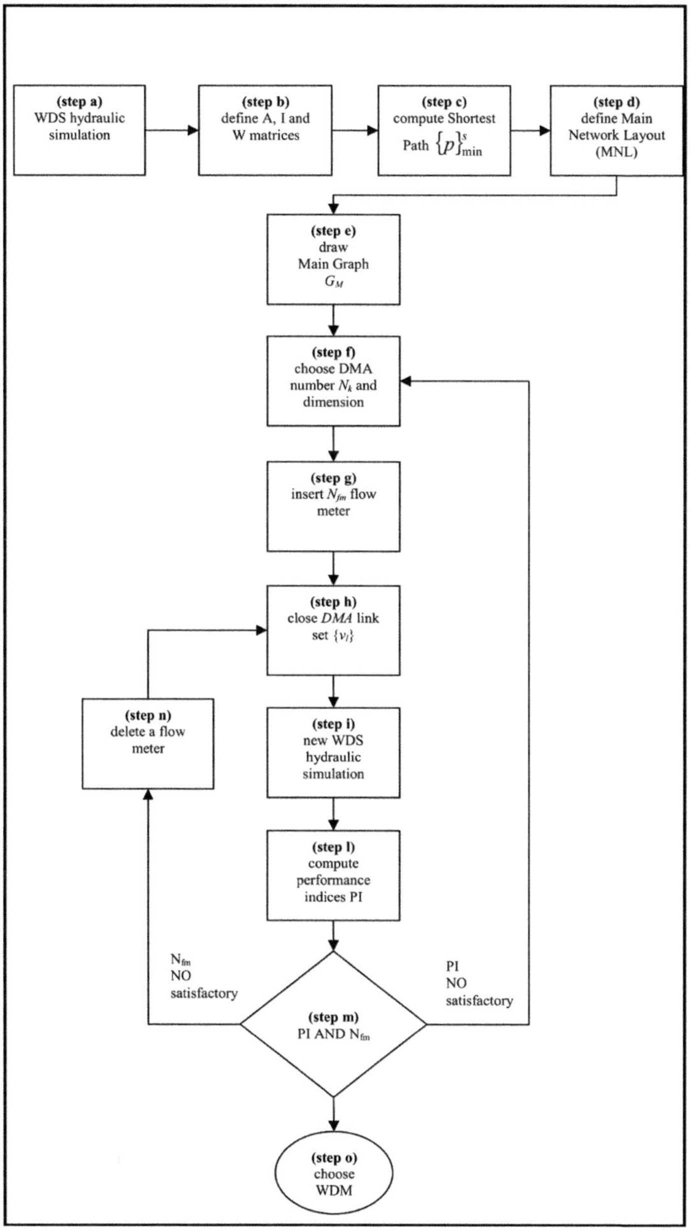

Figure 3. DSM flow chart

operating condition (peak water demand). Then it is easy to identify, by means of I matrix, pipe set $\{p_z\}^{s\,\beta_j} = \{\beta_s \ldots \beta_x, \ldots \beta_j\}^{s\,\beta_j}$ corresponding to node set $\{p_z\}^{s\,\alpha_i}$. Among all possible paths of set only one of them is the *minimum energy path* $p_{min}^{s\,\alpha_i} = \{\sigma_s, \ldots \alpha_x, \ldots \alpha_i\}^{s\,\alpha_i}{}_{min}$ with

$$\left(\min \sum_{\beta_j \in \{p_{min}\}^{s,\beta_j}} w_j \right)$$

calculated for corresponding path pipes.

Hence a new set $\{p\}_{min}^{s} = \{p_{min}{}^{s\,\alpha_i} \ldots p_{min}{}^{s\,\alpha_x}, \ldots p_{min}{}^{s\,\alpha_n}\}$ composed of all minimum energy paths **n** from reservoir s to each node α_i, can be defined (*step c*). This problem, known as the *Shortest Path Search*, in the case of a WDS can be solved by resorting to the graph theory determining the minimum cost path between two nodes (*s* and *i*) of a network directed graph G with *n* nodes, *m* links and a weight function w_j that correlates a cost ΔH_j to *j-th* pipes of the *G* [53] as in relation page 36. Many Shortest Path resolution algorithms [82] can be found in the literature; in this paper the Dijkstra algorithm was chosen [28] it allows all minimum energy paths $\{p\}_{min}^{s}$ to be easily obtained if the weight matrix *W* is known*(step c)*. Then, using adjacency matrix A', oriented according to flow direction, and defined as follows:

$$A'_{xy} = \begin{cases} 1 & \text{if } A'_{xy} = 1 \text{ and } W_{xy} > 0 \\ 0 & \text{else} \end{cases}$$

It is possible to draw the diagraph G of ONL by using different criteria of representation [23]. In the methodology proposed, a hierarchical approach was chosen [99, 42, 13, 85], in which all network nodes are drawn in different layers with a distinct hierarchy of connection.

This approach is a way of making hierarchies from diagraphs and of providing a good match between visual perception and network connection analysis identifying "ancestor" and "descendant" nodes of network layout defining a specific H_L hierarchical level [23].

It is thus possible to identify simpler specific structures as trees, loops, subsystems, groupings, etc., that can be needed during WDM design. In fact, in same cases a DMA can be defined by choosing all the network nodes that belong to a tree structure, or a specific subsystem, minimizing pipe cutting and monitoring water inflow more easily.

pipe counting how many times z_j each pipe j is found in all sn shortest paths obtained with the Dijkstra algorithm, as follows:

$$f_j = \frac{z_j}{sn} \tag{4}$$

In particular, pipes which are not included in any shortest path have f_j=0. Now it is easy to compute a new oriented adjacency matrix A" deleting all pipes with path frequency page 35 equal to zero and defined as follows:

$$A''_{xy} = \begin{cases} 1 & \text{if } (A_{xy} \neq\ 0) \text{ and } (f_y \neq\ 0) \\ 0 & \text{else} \end{cases}$$

So starting from A", a new network layout, called *Main Network Layout* (MNL), is defined *(step d)*, which differs from ONL because not all of its pipes are f_j=0. By means of matrix A" it is possible to draw the Main Graph G_M *(step e)* of the network with hierarchical representation that makes it possible to simplify system partitioning by helping planners choos the number and size of DMAs, as well as find the location of flow meters and boundary valves.

Three main statements govern the criteria adopted in the district Design Support Methodology to minimize the alterations of the performance of Original Network Layout (ONL):

1) The first, quite obviously, is that the higher the number of pipes closed to define DMAs, the higher the reduction in pipe diameter availability of the water supply system;
2) The second is that there are pipes in the network which can be defined as "less important" for system operations, meaning that they are present in the lowest number of total shortest paths. These pipes have a lower *path frequency* f_j as defined in page 35;
3) The third statement stems from the fact that, the Main Graph G_M being made by all the minimum energy paths which connect the reservoirs s to demand nodes *i*, the closure of one of its pipes forces the flow to change its path to reach the same demand node. Therefore, this alternative path, although compatible with the network topology, does not occur on the main graph and will be inevitably characterized by a higher head dissipation.

These observations lead us to define two criteria to be followed in order to select the pipes where flow meters and boundary valves are to be inserted for network WDM:

a) to minimize pipe closures and in particular those which belong to the Main Graph, by inserting flow meters on it preferably;

b) to reduce, if necessary, the number of flow meters by replacing them with boundary valves to be inserted in the pipes showing the lowest *path frequency*and cost w_j.

Therefore, once the number N_k and the size of districts (*step f*) have been defined according to the analysis of the Main Graph and to techno-economic considerations, it is possible to identify DMAs by inserting a certain number N_{f_m} of flow meters (*step g*) on the pipes of G_M.

After definition of the nodes of each district, uniquely determined by the position of the flow meters on G_M, it is easy - thanks to the incidence matrix I - to identify those pipes j (with f_j=0) connecting districts where N_{b_v} boundary valves are to be inserted in a pipe set $\{\nu_l\}$= $\{\nu_1 ... \nu_x, ... \nu_{N_{bv}}\}$ in order to obtain the permanent partitioning (*step h*). When the new configuration of the network is obtained, a new hydraulic test is performed (*step i*) and performance indices (PIs) are calculated *(step l)* in order to evaluate the WDM .

At this point, if PIs and N_{fm} are satisfactory *(step m)* for the planner the WDM is chosen (*step o*). On the contrary, should the PI not be satisfactory, it is necessary to reduce the number N_k or the size of DMAs, going back to *step f* since, obviously, the closure of pipes to create districts has caused an excessive increase in head losses which are not compatible with system performance.

But if PIs are satisfactory, it is possible to try and reduce the number N_{fm} of flow meters by replacing them, one by one, with boundary valves (*step n*), then reducing costs and simplifying the district water balance by using less flow meters. This operation can be carried out by removing, gradually, the flow meters inserted on the pipes of the Main Graph which show the lowest *path frequencies* and cost w_j according to criterion b).

Based on the above descriptions, DSM needs performance indices (PIs) in order to compare all the possible partitioning options of the water supply network and provide the planner with the necessary information to make the best choice. For this purpose several performance indices (PIs) are used, which allow the behaviour of a single district or the whole network to be tested. The indices, grouped into: (a) energetic, (b) statistical and (c) hydraulic are described below.

a) *Energy Indices*

According to [89], network looped topology allows for redundancy, which assists in ensuring that there is sufficient capacity in the system to overcome local failures and guarantee distribution of water to users.

The redundancy decrease can be measured with a "resilience" index which does not involve the statistical analysis of different types of uncertainty which are required for the definition of the reliability constraints.

Water District Metering can also be assimilated to permanent local failures that change the system layout by increasing internal energy dissipation and decreasing energetic redundancy; this effect is due to boundary valves that reduce network pipe diameter availability and remove some network loops.

Each WDM will tend to increase the internal energy dissipation and resilience indices can provide a good way to compare several system layouts. Specific resilience and failure indices were used starting from power balance of a water network [89], in the hypothesis of adiabatic pipes and of head loss $\Delta \mathrm{H}_j$ due to friction computed by a uniform flow formula:

$$P_A = P_D + P_N$$

in which each single equation term is represented by:

- Available Power (or total power):

$$P_A = \gamma \sum_{s=1}^{r} q_s H_s$$

where q_s and H_s are, respectively, the discharge and head relevant to each reservoir r and γ is the specific weight of water;

- Dissipated Power (or internal power):

$$P_D = \gamma \sum_{j=1}^{m} q_j \Delta H_j$$

where q_j and $\Delta \mathrm{H}_j$ are flow and head loss of each network pipe;

- Nodes Power (or external power):

$$P_N = \gamma \sum_{i=1}^{n} Q_i H_i$$

where Q_i and H_i are, respectively, water demand and pressure at each network node.

Then, similarly [89], two different indices are defined:

- Resilience index:

$$I_r = 1 - P_D/P_{D\max} \tag{5}$$

where

$$P_{D\ max} = P_A - \gamma \sum_{i=1}^{n} Q_i H_i$$

is the maximum power necessary to satisfy the demand constraints Q_i and node head constraint $H_i = e_i + h_i$ with h_i equal to the design pressure of the *i-th* node. Higher values of I_r page 38 indicate better WDMs in the sense of lower values of dissipated power and, hence, higher resilience;

- Resilience Deviation index:

$$I_{rD} = (1 - I_r^*/I_r)\ 100 = (P_D^* - P_D/P_{D\max} - P_D)\ 100 \tag{6}$$

where $I_r = P_D/P_{Dmax}$ is dissipated power of the WDM network layout. This index immediately shows immediately the resilience percentage deviation between WDM and ONL: naturally, higher values of page 38 mean worse WDM.

b) Statistical Indices

Energy indices refer to the entire water network but WDM also affects the individual DMAs. Therefore other district indices were used, specifically: district performance statistical indices that compute MEAN, MINIMUM, MAXIMUM and SQM values of hydraulic head for each district nodes.

Statistical indices allow us to summarize the most important information about nodal heads traditionally used to measure the level of service of WDS.

c) Hydraulic Indices

Finally district performance hydraulic indices were also developed in order to evaluate the behaviour of WDM with reference to district design pressure, as follows:

- Root of Mean Squares (RMS) of Pressure Deviations:

$$RMSPD_k = \sqrt{\frac{\sum_{i=l}^{n_k} \left(h_{ki} - \bar{h}_i\right)^2}{N_k}}$$

where h_{ki} is water pressure in the *i-th* node of the *k-th* DMA; h_i is design pressure in the same node; n_k is the number of the *k-th* district nodes. The $RMSPD_k$ index measures node pressure deviation, both positive and negative, from district design pressure at each node: low values of this index show a small alteration of DMA pressure distribution; otherwise high values indicate a strong effect of Water District Metering on original network status;

- Mean Pressure Surplus

$$MPS_k = \frac{\sum_{i=1}^{n_{Ok}} I_S}{n_{Sk}}$$

where

$$I_s = \begin{cases} 0 & \forall i: \bar{h}_i > h_i \\ h_i - \bar{h}_i & \forall i: \bar{h}_i \leq h_i \end{cases}$$

and n_{Sk} is the node number with $\bar{h}_i \geq h_i$ in the k-th district. This index gives a measure of node surplus: the higher the values, the higher the DMA pressure surplus and vice versa;

- Mean Pressure Deviations:

$$MPD_k = \frac{\sum_{i=1}^{n_{Ok}} I_D}{n_{Dk}}$$

where

$$I_D = \begin{cases} 0 & \forall i: h_{ki} > \bar{h}_i \\ h_i - \bar{h}_i & \forall i: h_{ki} \leq \bar{h}_i \end{cases}$$

and n_{Dk} is the node number with $h_{ki} \leq h_k$ in the *k-th* district. This index gives a measure of node pressure deviation: clearly the lower the negative values, the smaller the DMA pressure deviation and vice versa;

- Mean Frequency of Pressure Deviations:

$$MFPD_{kx} = \frac{\sum_{i=1}^{n_{xk}} I_{DFx}}{n_{xk}}$$

where

$$I_{DFx} = \begin{cases} 0 & \forall i : \bar{h}_{ki} > \bar{h}_i - x \\ 1 & \forall i : \bar{h}_{ki} \leq \bar{h}_i - x \end{cases}$$

and h_{xk} is the node number of the k-th district in which pressure is x meters less than node design pressure $\bar{h}_i$. The MFPD$_{kx}$ index measures the frequency of pressure deviation from prefixed pressure values $\bar{h}_i - x$ in order to have a more specific knowledge of pressure deviation distribution.

In the paper the value of x was chosen as equal to 0, 2 and 5 m, in order to test the pressure deviation range; in general, low values of this index show good system performance but, more specifically, high MFPD$_{k0}$ values can also indicate very small deviations from the district design pressure while low MFPD$_{k10}$ indicates large pressure deviation in DMAs.

3.2 Water Pressure Management Design

As already indicated in section 1.2, pressure management has to take into account the variation of demand at different times; typically users' demand is almost zero at night and is maximum at peak hours.

Furthermore, due to the degree of meshing which generally characterizes household distribution networks, pressure adjustment inside a single district can affect the operation of neighbouring districts, with a kind of interaction that is not linear because of the known characteristics of turbulent motion.

In order to solve this kind of problem, a specific optimization procedure, based on genetic algorithms is implemented by guaranteeing its compatibility with a widespread "open source" calculation code for pressurized water supply networks. This approach is based on the simultaneous identification of hourly demand and leakage pattern performed using time histories of pressure and flow at a few sites on the network, with the aid of genetic algorithms in the hypothesis that:

a) water consumption by users independent of pressure;
b) homogeneous distribution of per-capita daily freshwater demand;
c) homogeneous values of the parameters in the pressure-leakage law; discharge of the i-th node on the k-th hour is evaluated as:

$$D_i^k = n_{ui}\, c_k^d + a\left(P_i^k\right)^b$$

where in the first right-hand term of the equation the $n_{u,i}$ is the number of users connected to the i-th node and c_k^d is the demand coefficient of the k-th hour, whose daily average represents the daily per-capita freshwater demand. The second term on the right-hand side is clearly equation page 12 written with reference to the k-*th* hour.

The validity of the combined Demand-Leakage model page 41 for the realistic modeling of the water supply network has been widely recognised (for instance [105]). Due to the dependence on pressure of the second addendum in page 41, a hybrid Demand/Pressure-Driven method is needed for network hydraulic simulation. The procedure is arranged in some different steps illustrated in the following flow chart of Figure 3:

a) identify Demand-Leakage law $\{c_k^d$,a,b$\}$'
b) define PRV optimal setpoint;
c) operate pressure management;
d) identify Demand-Leakage law again $\{c_k^d$,a,b$\}$"
e) if the difference between $\{c_k^d$,a,b$\}$ and $\{c_k^d$,a,b$\}$ is minor ϵ of choose PRV setpoint;
f) else return to step b) and define the PRV optimal setpoint again.

As already cited in section 1.2, in order to identify the Demand-Leakage law page 12 the methodology presented in [27] was used.

Starting from a mathematical model of the network, the values of the three parameters of model page 41, allowing the best reproduction of measured pressures and flows, based on experimental n_{m_F} and n_{m_P} available time series of n_h measures of hourly averaged flow and pressure, respectively, can be found by solving a properly defined optimization problem.

For any given set of parameter values $n_{u,i}$ a and b, once the hydraulic simulation of the network has been performed, the minimization of the following Objective Function (OF) can be computed :

$$\min_{c_k^d,a,b} (OF) \tag{7}$$

$$OF = \frac{\sqrt{\sum_{k=1}^{n_h}\sum_{i=1}^{n_{mF}}\left(\frac{Q_{C,i}-Q_{M,i}}{Q_{M,i}}\right)^2+\sum_{i=1}^{n_{mP}}\left(\frac{P_{C,i}-P_{M,i}}{P_{M,i}}\right)^2}}{\sqrt{n_h}(n_{mF}+n_{mP})}$$

Since the OF represents a measure of the deviation of computed values compared with the observed ones, the optimal choice of the above parameters set is naturally defined as that minimizing the Objective Function. The search of the above minimum is performed through an Enhanced Genetic Algorithm, implemented with MATLAB (by The MathWorks).

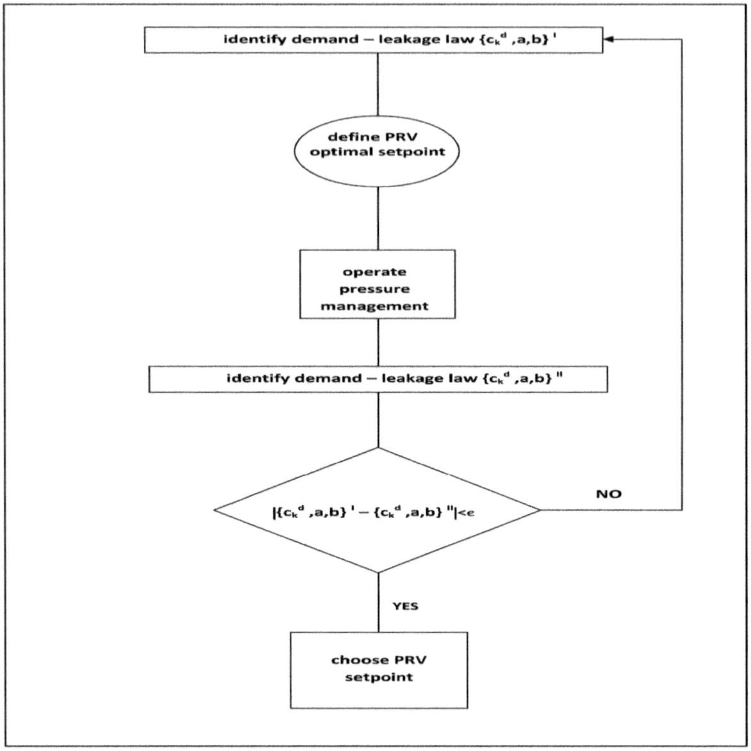

Figure 4. Flow Chart of Water Pressure Managment Design

4. Case study: Monterusciello network

As in many other countries also in Italy the leakage problem has been long known [63, 61].For the purpose of damage limitation, the Ministerial Decree for Public Works no. 99 was issued on 8 January 1997, which represents the first, reliable step, to deal with the control of wastes in the management of water resources, with special focus on urban distribution networks [5].

Furthermore, with regards to the national context, the IWA scheme can be substantially superimposed, by identifying the relevant similarities, on that indicated by Italian legislation in M.D. no.99 of 8 January 1997 [5].

In this chapter a case study is illustrated, namely WDM of the pilot site of Monterusciello, initiated by the Interdipartimental Research Center in Environmental Engineering (CIRIAM) of the Second University of Napoles (SUN) in 2006 and developed within the HYDRANET European Project in the framework of the INTERREG IIIB/MEDOCC which provides for transnational and interregional cooperation between several European nations and the countries of the southern shores of the Mediterranean Sea, in order to promote their social and economic development and strengthen their cooperation with the EU countries through the exchange of technical and scientific knowledge.

Greece was the leader of the HYDRANET project (with Rodopi Development and the University of Democritus in Thrace), while the other partners are: Italy (with the Second University di Napoles and Regional Authority of Tuscany), France (with the Universit de Marseille), Morocco (with two research laboratories), Tunisia (with the National School of Engineering at Monastir) and Lebanon (with the Laboratory of Hydraulic Systems of the School of Agriculture).

Specifically, HYDRANET aimed at contributing to spreading the concepts of analysis and monitoring of the hydrological cycle of the Mediterranean Basin, by developing control systems for saving, recycling and reusing water and by defining new integrated water resource management plans. HYDRANET sought to develop proposals for planning pilot projects, by testing hydrological models and management method, for example to reduce waste in the water supply networks for civil and agricultural use, to combat desertification, and to control the salinity and quality of water in soils, etc..

In particular, the SUN had the task of analysing - through the creation of a pilot site - the effectiveness of district metering both to evaluate the district water outcomes and to adjust pressure in the distribution networks in order to reduce water leakages.

4.1 Pilot site description

The study-area is the district of Monterusciello in Pozzuoli. It covers an area of about 100ha, has a maximum altitude of 126.60m and minimum of 24m. These two altitudes are 1.5km away; the maximum slope being about 7%, this area can be considered as rather flat, with an average altitude of 75m a.s.l.

Monterusciello town-planning policies were deeply affected by the episodes of bradyseism in 1983. In that period bradyseism was particularly intense; the panel of experts cooperating with the Ministry of Civil Protection decided to divide the Pozzuoli built-up area into two zones: zone "A", which included the old city centre, via Napoli (the road leading to the city of Naples), the port and the Solfatara (an area of volcanic craters), was identified as a "high risk" zone inasmuch as it was likely to be the epicentre of a perspective strong seismic event and zone "B" which would be directly concerned with the seism if the epicentre were not located in zone "A".

A decision was made that all the residents in zone A would have to move and that a new town would be built (Monterusciello 2) in the shortest possible term, under government licence, in the former Italsider site besides guaranteeing rapid completion of the construction of Monterusciello 1 which had already started years before.

The agreement on Monterusciello 2 between the Ministry of Civil Protection and the University "Federico II" of Naples was signed in 1983; the planning stage was over by early 1984. Six hundred prefabricated houses were built in Monterusciello 1, whose development had originally started in the framework of Council Houses (Law no. 167/62); the site was split into a west-side area for cooperative societies (traditional buildings) and an east-side area. In Monterusciello 2, a much larger area, some heavyweight prefabricated structures were built between 1984 and 1987.

The interventions for new residential buildings at Monterusciello 2 provided for the construction of about 4,500 flats in 267 buildings, grouped into 19 building lots, for a population of about 16,000 inhabitants, with a population density of 3.56 inhabitants/flat.

The whole network has been mapped within CAD and GIS environments for data analysis and representation purposes. The area of Monterusciello involved in the study is shown in Figure 5.

In the framework of the Hydranet project, some innovative equipment

Figure 5. Pilot Site of Monterusciello 2 (Google-Tele Atlas 2008)

to measure and monitor water leakages within the network was installed in this pilot site. Currently, the situation is as follows:

- there are several areas where public spaces should have been landscaped and furnished which have been left as wasteland;
- some infrastructures are not utilised at present and are deteriorating (such as shopping malls); the infrastructures which are working are in very poor maintenance conditions;
- the residential buildings, which were built hurriedly and with inadequate technologies, are in poor conditions due to many co-factors: poor quality of materials, lack of regular maintenance (in most cases maintenance interventions are performed occasionally, only when there

is an emergency), improper use of the dwellings by assignees, such as the addition of new elements altering the structure;

- the obsolescence of the technical plants which have not been modified in compliance with the new regulations enforced over the years.

This is the context of the study of the water supply network located in the Monterusciello 2 area, as shown in Figure 5.

In the framework of the EU-funded research project Hydranet, CIRIAM set out to test the application of modern water system remote monitoring devices in order to evaluate possible benefits on district networks.

It is worth mentioning that in Italy the provisions governing the creation of district meter areas was only recently enforced by Ministerial Decree n. 99 dated 8 January 1997 that, as concerns district metering, provides that "*[...] To control leakages in a water distribution system it is useful to split the plant into districts [...]*, and gives to this word the following meaning, "*[...] portions of a water distribution network for which a fixed metering system has been installed to calculate the volume of water inflow and outflow [...]*", it is possible to assume two kinds of district metering, according to different techniques of water distribution network partitioning.

Nevertheless, due to the peculiarities of the Italian system as compared to other European realities, especially in terms of management deficiencies, and in consideration of the technological breakthroughs in the field of remote monitoring, CIRIAM decided to test districts and the management of pressure within a water supply network which is large enough to represent an actual scale network both modelling and management wise.

As mentioned above, Monterusciello 2 was built in order to provide dwellings to people who were forced to leave Pozzuoli due to bradyseism. Therefore, the site is intended only for residential activities and a few fundamental services. In particular, in addition to about 4,500 flats, there are:

- a church, a bank, a post office;
- a Town Hall;
- a high school: scientific studies;
- 3 schools: a primary school, a secondary school, a vocational school for the hotel industry;
- a junior high school;
- a church, a school, a Carabinieri station (police station), a Fire Brigade, an Electric Company (ENEL) office, an indoor sports arena with a swimming pool.

Water consumption is mostly for household use with a negligible percentage for other uses. There are neither major public fountains nor important road cleaning activities.

In order to calibrate the simulation model, some criteria were used for the collection and spatial distribution of data to calculate households and relevant consumption. Photographic surveys and inspections were made easier by the presence of repetitive building types.

The number of regular contracts with the National Energy Authority(ENEL) was compared with the number of flats and/or shops in the area using a value of 3.56 inhab./flat (reported by the census of the Pozzuoli municipality).

Since ENEL data were aggregated with a higher complexity, the data used for the spatial distribution were those obtained from the analysis on the ground.

A total resident population of 15,689 people was thus adopted with an optimal agreement between ENEL data and estimated data.

4.2 The hydraulic model

Hydraulic software used to model and simulate the Monterusciello network was EPANET 2 [75], developed by the National Risk Management Research Laboratory of the Environmental Protection Agency(EPA) of the USA.

EPANET is a software programme that performs extended period simulation of hydraulic and water quality behaviour within pressurized pipe networks.

A network consists of pipes, nodes (pipe junctions), pumps, valves and storage tanks or reservoirs. EPANET tracks the flow of water in each pipe, the pressure at each node, the height of water in each tank, and the concentration of a chemical species throughout the network during a simulation period comprised of multiple time steps.

In addition to chemical species, water age and source tracing can also be simulated. EPANET is designed to be a research tool for improving the understanding of the movement and fate of drinking water constituents within distribution systems. It can be used for many different kinds of applications in distribution systems analysis. Sampling program design, hydraulic model calibration, chlorine residual analysis, and consumer exposure assessment are some examples.

EPANET can help assess alternative management strategies for improving water quality throughout a system. Running under Windows, EPANET provides an integrated environment for editing network input data, running hydraulic and water quality simulations, and viewing the results in a variety of formats.

The Monterusciello water supply network has 120 nodes, 150 interconnection branches and 7 antennas; there are 72 manholes of various sizes

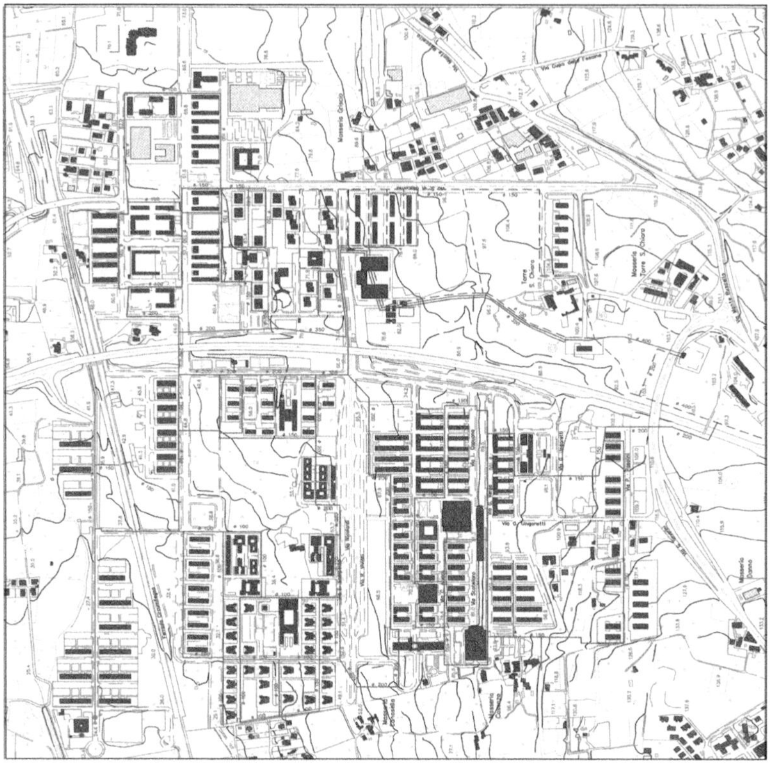

Figure 6. Water network of Monterusciello 2

(Figure 6).

The network is fed by two reservoir: one located at 161.40m a.s.l., which supplies part of the properties at high elevations, the other located at 104.20m a.s.l. for properties with low elevations. The network spreads for about 23km and is made of steel pipes with a diameter ranging between DN500 for adduction and DN100 for distribution.

Based on the building type, as stated above, it may be assumed the water supply network is homogeneous in terms of material (steel) and age (about 20 years). Furthermore, the kind of prefabricated flats allows the scheme of connections to be simplified as to create a water supply model of the system.

The hydraulic parameters of the Monterusciello networks are shown in Table 2s.

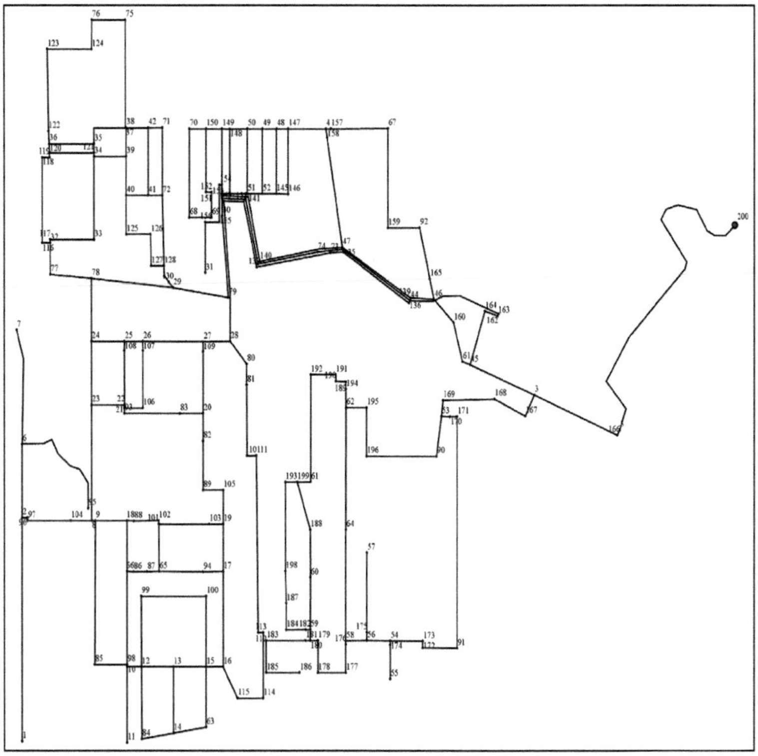

Figure 7. The hydraulic model of Monterusciello

nodes (n)	200
pipes (m)	231
total length (m)	23.5
pipes material	steel
reservoir (node 200)	1(161.40 m)
roughness (Manning)	0.015
peak total flow (l/s)	90.8

Table 2. Monterusciello network characteristics

4.3 District Metering design

Each single step of DSM application, illustrated in flow chart of Figure 4 of Chapter 3, was applied to the case study of the Monterusciello 2 network (Figure 7).

Design pressure for all nodes was assumed equal to $h_i = h = 28m\ \forall i = 1..n$ compatible with building height.

Following the DSM flow chart reported in Figure 3, the first *step a)* was hydraulic simulation of the Monterusciello network model, reported in Figure 7 and Table 2, by means of EPANET software in order to calculate unknown pipe flows q_j and nodal heads h_i. Hydraulic simulation was carried out during peak water demand because permanent WDM effects on hydraulic performance are clearly worse in this operating condition.

Then (*step b*) adjacency *A*, incidence *I*, weight *W* and oriented adjacency *A'* matrices were defined and, subsequently (*step c*) all shortest paths $\{p\}^s_{min}$ from each reservoir to all network nodes were computed by the Dijkstra algorithm.

For example in Table 3, a few minimum paths of the Monterusciello network, from the reservoir $\sigma_1 = 200$ to the i-th node, are represented.

After all path frequencies f_j had been calculated (*step c*) it was possible to define a new MNL (step d) computing matrix A" (10); specifically there are 32 Monterusciello network pipes with f_j=0.

Then, thanks to the MATLAB Bioinformatics Toolbox (by The MathWorks),the oriented graph G (Figure 8) was drawn and, by adopting the hierarchical criterion to represent nodes [81], also the oriented Main Graph G_M (Figure 9) of MNL have been drawn. In particular, Figure 9 also shows the H_L=32 hierarchical levels into which nodes are divided.

By comparing the two figures, it is possible to see that, unlike graph G (Figure 8), the Main Graph G_M (Figure 9) network topology is more open or tree-like which makes the district design easier.

On comparing the two graphs in Figure 8 and 9 with the traditional layout of the water network in Figure 7 it is easy to realize that this visualization is more effective to design a WDM.

In particular, the main graph G_M (Figure 9) allows immediate visualization of the "most important" pipes of the water system which generally correspond to those with a lower hierarchical level H_L and a higher path frequency. Furthermore, the G_M allows an easier visualization of the several node groups of the network which can give useful information to choose the number and size of districts.

Therefore, based on the analysis of the Main Graph, DSM suggested partitioning the network into $N_k = 7$ DMAs (*step f*) by inserting $N_{f_m} = 7$

Shortest path set	Shortest path	Minimum energy	Label α_i of shortest path nodes
$\{p\}_{\min}^{200}$	$p_{\min}^{200,\alpha_1}$	$\sum_{\beta_j \in \{p_{\min}\}^{s,\alpha_1}} w_j$	$\sigma_1,\ldots,\alpha_1$
	$p_{\min}^{200,\alpha_2}$	60,84	200,5,166,3,45,164,163,199,162,44,136,135
	$p_{\min}^{200,\alpha_3}$	60,70	200,5,166,3,45,164,163,199,162,44,136
	$p_{\min}^{200,\alpha_4}$	60,99	200,5,166,3,45,164,163,199,162,44,136,135,73,137
	$p_{\min}^{200,\alpha_5}$	61,09	200,5,166,3,45,164,163,199,162,44,136,135,73,137,138
	⋮	⋮	...
	$p_{\min}^{200,\alpha_x}$	$\sum_{\beta_j \in \{p_{\min}\}^{s,\alpha_x}} w_j$	$\sigma_1,\ldots,\alpha_x$
	⋮	⋮	...
	$p_{\min}^{200,\alpha_n}$	$\sum_{\beta_j \in \{p_{\min}\}^{s,\alpha_n}} w_j$	$\sigma_1,\ldots,\alpha_n$

Table 3. Monterusciello network shortest paths from reservoir $\sigma_1 = 200$ to all α_i nodes

flow meters (*step g*) which, along with the flow meters traditionally installed downstream of the reservoir, will allow district water balances to be made in order to evaluate the flow losses in accordance with the sizes of DMAs suggested in the literature.

At this point, once the individual nodes belonging to each DMA were known, it was easy to identify all connection pipes j in which to insert the $N_{b_v} = 7$ boundary valves (*step h*) in order to obtain the 7 DMAs uniquely identifying the WDM ID 1, shown in Figures 9. The single pipes j (with the relevant f_j), where flow meters and boundary valves were inserted, are reported in the first row of Table 4, together with the total number of nodes of each DMA.

A new simulation of the water network was then performed (*step i*) and PIs were calculated (*step l*), as defined in section 3.2, whose results are reported in Tables 4, 5, 6 and 7 corresponding to WDM ID 1.

Subsequently, since the PIs can be considered satisfactory, it was decided

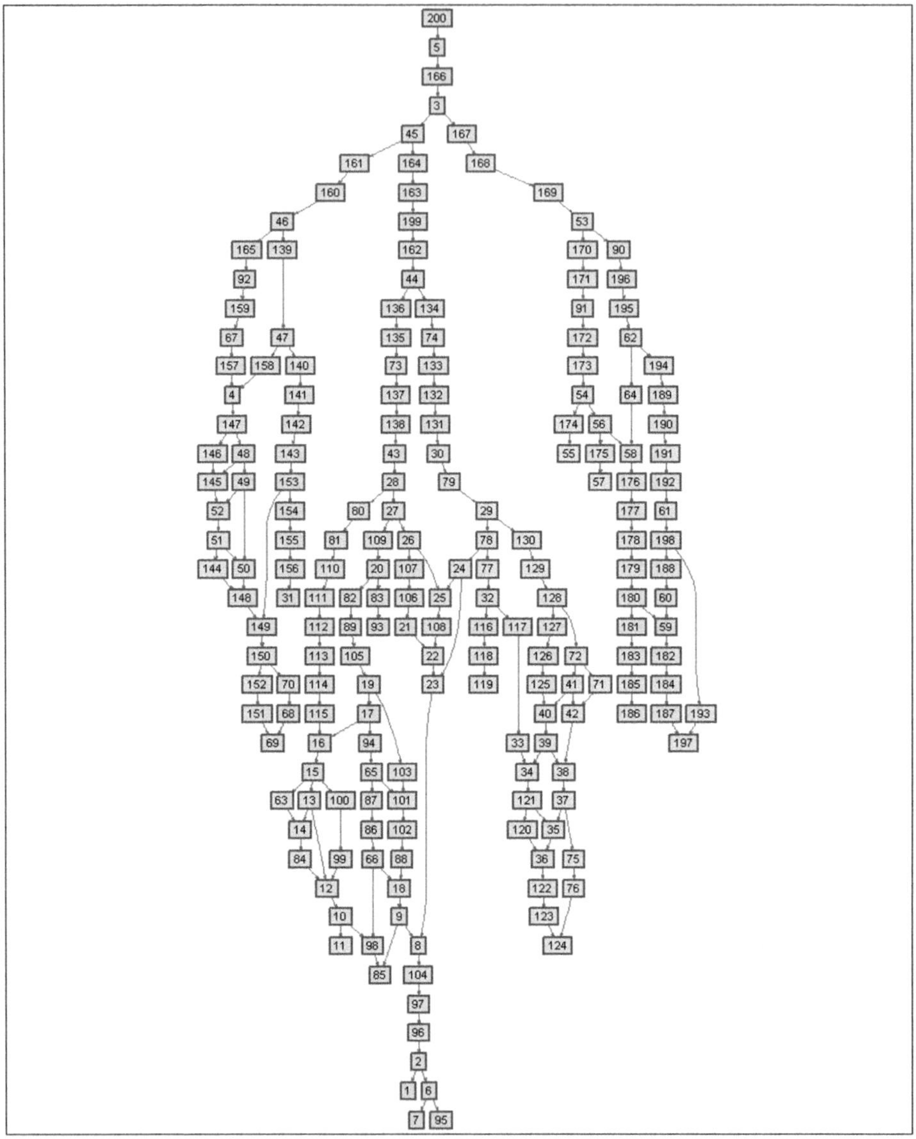

Figure 8. Graph G of the Monterusciello network

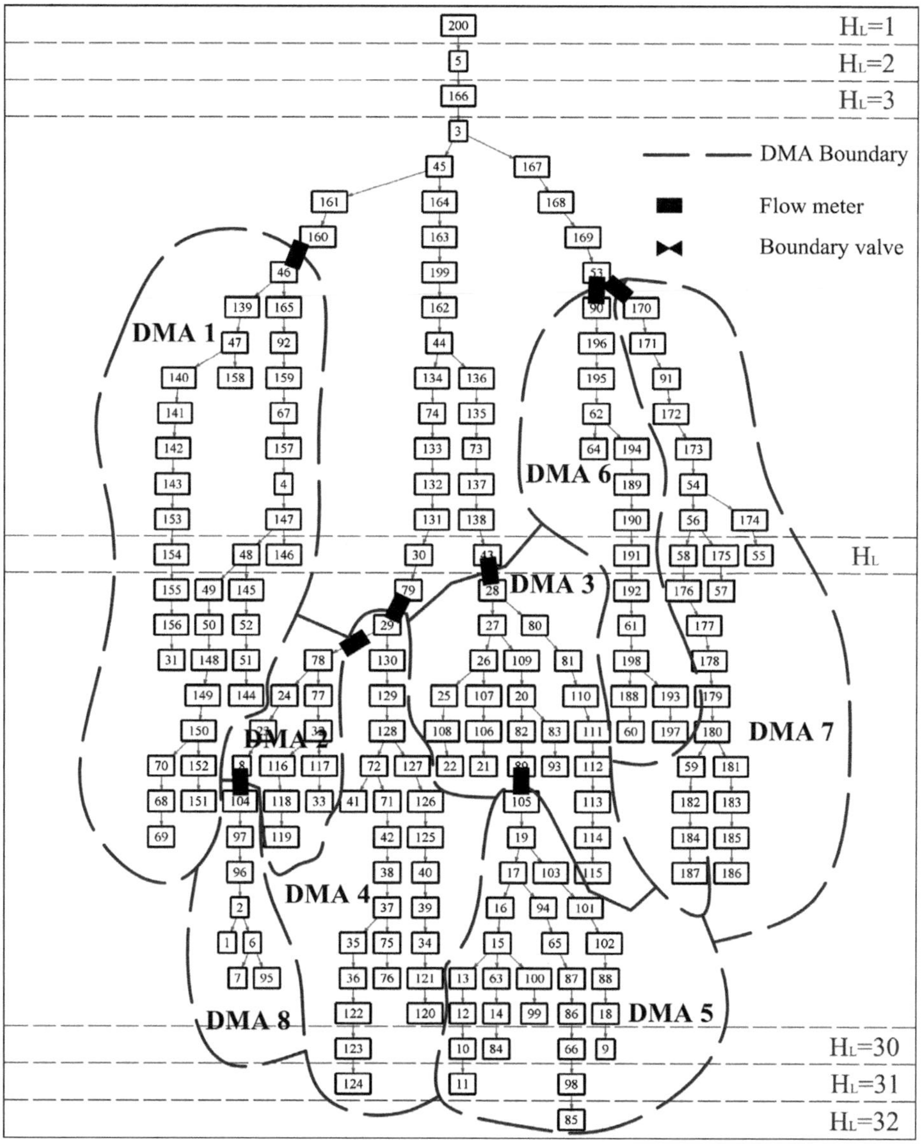

Figure 9. Main Graph G_M of the Monterusciello network showing of 7 DMAs and 8 flow meters (WDM ID 1)

to analyze three other WDM layouts each with 7 DMA one, as reported in Tables 4, 5, and 7 (identified with ID labels 2, 3 and 4).

Specifically WDM ID 2 has $N_{fm} = 9$ and $N_{bv} = 4$, WDM ID 3 has $N_{fm}=8$ and $N_{bv}=5$, WDM ID 4 has $N_{fm}=9$ and $N_{bv}=5$. Each of them was created by straightforwardly modifying WDM ID 1 according to criterion b) (see section 3.1), by inserting boundary valves in pipe j of G_M with the lowest path frequencies f_j, and to design constraints (in the Monterusciello case according to the position of existing boundary valves).

Then *step h*) was repeated several times, creating three other WDMs with a different number of boundary valves and flow meters.

The results of simulations, for each WDM, were reported in Tables 4, 5, 6 and 7, as a function of the sum F of all the path frequencies f_j of pipes β_j belonging to the set $\{\nu_l\}$ of the N_{bv} boundary valves closed to define districts:

$$F = \sum_{\beta_j \in \{\nu_l\}} f_j \tag{8}$$

First of all, by analyzing *Table 5*, it is worth noticing that the Original Network Layout (ONL) shows a rather low value of the resilience index $I_r=0.392$ which confirms the low redundancy of the system with a design pressure *h=28m*. It can also be observed that each WDM obtained with DSM shows a good I_{rd} value.

In particular, the WDM A1 shows a value *F=0* since it was created without closing any pipes on G_M; indeed, in this case, the only pipes closed by boundary valves are not included in any shortest path and they all have, obviously, $f_j=0$.

The effectiveness of the proposed approach can be evaluated by observing that the four WDMs (ID 1, ID 2, ID 3 and ID 4) show low deviations from the resilience index, below 2.10, with a negligible increase in dissipated power P_D and equal to 38 kWatt at the maximum. Such small deviations are compatible with the performances of the system as confirmed by statistical (*Table 6*) and hydraulic (*Table 7*) indices.

To be more precise, in order to analyze performances in terms of node hydraulic head, statistical indices of the Monterusciello network indicate good results in each DMA of all WDM ID 1, ID 2, ID 3 and ID 4.

Furthermore, the statistical indices allow to point out how the WDM alters the distribution of network pressures in a non-homogeneous way.

Indeed, *Table 6* shows, for example, with reference to WDM ID 1 an increase in the mean pressure in DMAs 4 and 7 and a decrease in DMAs 2, 3, 5 and 6; the same occurs for minimum and maximum pressures with relatively reduced SQM. Such system behaviour is due both to the layout changes made with WDMs and the position of the reservoir which may, in some cases, increase the pressure of a district to the detriment of others. It is therefore advisable to use indices other than the resilience one which, alone, cannot show the effect of the WDM on individual DMAs.

Also hydraulic indices, reported in *Table 7*, confirm the excellent performances of WDM ID 1 obtained with DSM. The $RSMPD_k$ (17) index shows that, vis-à-vis the design pressure h, there is a slight worsening of the performance of DMA 2, 3 and 5, whereas the worsening is stronger for DMA 2 and 5 in WDM ID 2, ID 3 and ID 4. In DMA 3 of WDM ID 1, as already pointed out with statistical indices, there is a slight improvement that results into a decrease in the deviation, measured by MPD_k (20), in those nodes with pressure lower than the design value, and in an increase in the surplus measured by MPS_k (18), in those nodes with pressure higher than h. By contrast in DMA 7, MPD_k assumes negative values in WDM ID 2, ID 3 and ID 4 and there is a slight worsening of the surplus index MPS_k.

We may further evaluate the analysis of the performance of single districts by using the $MFPD_{kx}$ (22) index, which measures the mean frequency of the deviation of single nodes. In all cases the $MFPD_{kx}$, calculated for x=0, 2 and 5 m, shows that no one node is below pressure except for some minor exceptions.

WDM				DMA Boundary Pipes			DMA number nodes						
N_k	ID	N_{fm}	N_{bv}	Flow Meter in G_M	Boundary Valves in G_M (with relative f_j)	Boundary Valves out G_M (with f_j=0)	1	2	3	4	5	6	7
7	1	7	9	179,63,133,36,96,76,190	-	20,23,48,115, 118, 225,227,228,229	36	19	27	23	27	16	25
	2	9	4	179,64,226,26,66,68,222,76,190	210(1)	225,227,229	37	8	23	29	33	15	26
	3	9	5	132,63,68,179,66,26,190,76,222	210(1), 61(6)	225,227,229	37	8	26	29	30	15	26
	4	8	5	179,64, 26,66,68,222,76,190	210(1), 226(11)	225,227,229	37	8	23	29	33	15	26

Table 4. Water District Metering Layouts with 7 DMAs

This result confirms the effectiveness of DSM since it is clear that a decrease in pressure by a few meters only is acceptable for the planner as regards the goals which can be achieved through partitioning.

The results achieved for the WDM in 7 districts show that the DSM gives the planner several possible solutions compatibly with design constraints.

A possible choice could be indifferently made on WDM ID 1, ID 2, ID 3 or ID 4 which show a slightly higher deviation for I_{rd}. Each choice would be reasonable when evaluating district statistical and hydraulic indices of each WDM which can be considered, in any case, satisfactory.

WDM ID	Power (kWatt)			Energy Indices		F
	P_A	P_N	P_D	I_r	I_{rd}	
ONL	143.8	106.3	37.5	0.392	-	-
1	143.8	106.1	37.7	0.388	1.02	0
2	143.8	106.0	37.8	0.388	1.02	1
3	143.8	105.8	38	0.384	2.05	7
4	143.8	105.9	37.9	0.386	2.10	12

Table 5. Energy Indices of Monterusciello network WDMs and ONL)

WDM ID	INDEX	DMA 1	DMA 2	DMA 3	DMA 4	DMA 5	DMA 6	DMA 7	INDEX	DMA 1	DMA 2	DMA 3	DMA 4	DMA 5	DMA 6	DMA 7
ONL	MEAN	74.08	54.23	34.60	40.22	57.51	70.56	64.21	MIN	51.45	38.91	24.81	30.97	45.42	54.55	28.27
1		74.08	53.55	34.41	40.49	56.78	70.35	64.36		51.45	38.52	24.71	31.18	45.06	54.53	28.35
2		73.88	33.27	52.10	55.69	37.93	70.29	63.94		51.45	31.06	37.62	42.47	28.08	54.66	27.71
3		73.88	33.31	51.08	55.76	36.05	70.29	63.94		51.45	31.11	33.61	42.55	26.77	54.66	27.71
4		73.88	32.98	51.58	55.22	38.22	70.29	63.94		51.45	30.78	37.30	42.00	28.33	54.66	27.71
ONL	MAX	89.73	74.16	43.29	50.94	68.72	82.59	89.37	SQM	12.72	8.98	4.67	7.09	7.10	7.95	16.83
1		89.73	73.15	43.06	51.12	67.93	82.16	89.57		12.72	8.69	4.65	6.97	7.01	7.84	16.87
2		89.73	39.41	74.04	67.71	55.03	82.39	88.66		12.60	3.09	9.41	7.51	6.33	7.94	16.58
3		89.73	39.46	74.13	67.78	46.66	82.39	88.66		12.60	3.09	9.89	7.51	5.48	7.94	16.58
4		89.73	39.13	73.50	67.24	55.51	82.39	88.66		12.60	3.09	9.37	7.51	6.37	7.94	16.58

Table 6. Statistical Indices of Monterusciello network WDMs and ONL

WDM ID	INDEX	DMA 1	DMA 2	DMA 3	DMA 4	DMA 5	DMA 6	DMA 7	INDEX	DMA 1	DMA 2	DMA 3	DMA 4	DMA 5	DMA 6	DMA 7
NDM	RMSPD	47.76	27.65	8.03	14.05	30.32	43.25	39.79	MPS	46.08	26.23	7.32	12.22	29.51	42.56	36.21
1		47.76	26.91	7.87	14.23	29.59	43.03	39.94		46.08	25.55	7.43	12.49	28.78	42.35	36.36
2		47.54	6.01	25.79	28.66	11.73	42.98	39.45		45.88	5.27	24.10	27.69	9.93	42.29	37.39
3		47.54	6.05	25.03	28.73	9.69	42.98	39.45		45.88	5.31	23.08	27.76	9.06	42.29	37.39
4		47.54	5.76	25.29	28.20	11.99	42.98	39.45		45.88	4.98	23.58	27.22	10.22	42.29	37.39
NDM	MPD	0.00	0.00	-2.43	0.00	0.00	0.00	0.00	$MFPD_0$	0.00	0.00	0.07	0.00	0.00	0.00	0.00
1		0.00	0.00	-1.72	0.00	0.00	0.00	0.00		0.00	0.00	0.11	0.00	0.00	0.00	0.00
2		0.00	0.00	0.00	0.00	0.00	0.00	-0.29		0.00	0.00	0.00	0.00	0.00	0.00	0.04
3		0.00	0.00	0.00	0.00	-1.04	0.00	-0.29		0.00	0.00	0.00	0.00	0.00	0.00	0.04
4		0.00	0.00	0.00	0.00	0.00	0.00	-0.29		0.00	0.00	0.00	0.00	0.10	0.00	0.04
NDM	$MFPD_2$	0.00	0.00	0.04	0.00	0.00	0.00	0.00	$MFPD_5$	0.00	0.00	0.00	0.00	0.00	0.00	0.00
1		0.00	0.00	0.04	0.00	0.00	0.00	0.00		0.00	0.00	0.00	0.00	0.00	0.00	0.00
2		0.00	0.00	0.00	0.00	0.00	0.00	0.00		0.00	0.00	0.00	0.00	0.00	0.00	0.00
3		0.00	0.00	0.00	0.00	0.00	0.00	0.00		0.00	0.00	0.00	0.00	0.00	0.00	0.00
4		0.00	0.00	0.00	0.00	0.00	0.00	0.00		0.00	0.00	0.00	0.00	0.00	0.00	0.00

Table 7. Hydraulic Indices of Monterusciello network WDMs and ONL

In Figure 10a, a Main Graph of WDM ID 2 was reported, in which a valve of G_M was inserted, differently from WDM ID 1 (Figure 9), and in Figure 10b WDM ID 2 was reported directly on the traditional representation of the hydraulic model of the Monterusciello network.

The creation of districts complies with the provisions of M.D. no. 99/1997 which, however, is aligned with the main indications of the technical-scientific literature and international regulations. Therefore, this choice is no limit to the research work carried out at Monterusciello.

Therefore, starting from results obtained with DSM, seven permanent water supply districts were planned, called D.M.A.s (District Meter Areas) A1, A2, B1, B2, B3, C, D as represented in Figure 11.

Based on the above evaluations, seven districts were designed whose main characteristics are listed in the following tables (Tab. 8, 9, 10, 11, 12, 13, 14, 15).

4.4 Measurement devices

The devices selected for districts are modern integrated systems for metering and monitoring the main hydraulic (flow and pressure) and operational (battery level) parameters. They can be divided into two groups:

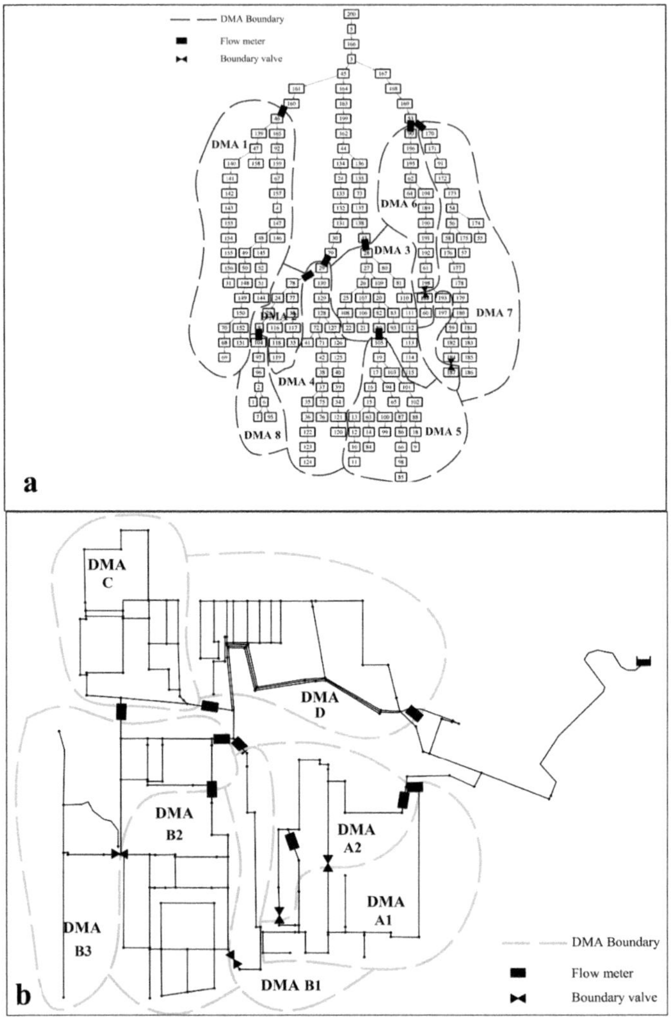

Figure 10. Traditional representation of Monterusciello WDM A2

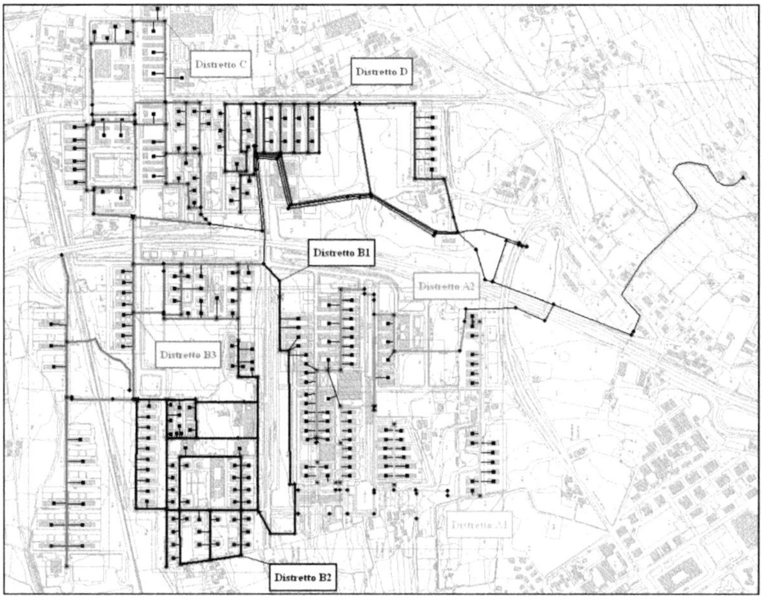

Figure 11. District metering project of the water distribution network of Monterusciello

	DMA A1	DMA A2	DMA B1	DMA B2	DMA B3	DMA C	DMA D
District size (ha)	20.90	8.87	9.23	18.37	27.68	34.37	15.40
N° of buildings	50	24	16	50	35	61	26
N° of estimated users	780	544	308	730	908	994	424
N° of ENEL users	764	539	325	720	909	1.022	443
Estimated inhabitants	2,777	1,937	1,096	2,599	2,232	3,539	1,509
Portions of water network	15	7	6	28	18	38	16
Length of the network (m)	2,550	1,400	920	3,600	3,100	3,650	3,950
No. of nodes	21	11	7	22	13	31	31
Max height (m a.s.l.)	126.60	105.60	68.80	59.30	66.00	74.00	106.4
Min, height (m a.s.l.)	70.00	70.00	52.10	30.00	24.00	44.80	71.80
Pressure level	AP	AP	BP	BP	BP	BP	BP

Table 8. Summary of data for the seven districts in Monterusciello

District extension	**20,90 ha**				
Building type	Prefabricated				Traditional structure
Photos	PHOTO 1 	PHOTO 3 	PHOTO 2 		PHOTO 4 
Buildings	16	8	4	8	14
Floors	5	4	4	4	-
Apartment for floor	6	5	4 (2 buildings) 3 (2 buildings)	6	-
Residential floors	1	1	1	-	-
Commercial floors	-	-	-	-	-
Stores that need water	-	-	-	-	-
Public buildings	1 church, 1 bank, 1 postal office				
Users for type	384	120	186		60
Users	780*				
Enel users	764				
Inhabitants	2777				

	Pipes	*Diameter*	*Length [m]*	*Nodes*	*Height max [m]*	*Height min [m]*	*Roughness*
Hydraulic parameters	10	150	1908,20	21	126,60	70,00	-
	5	100	636,52				-
TOT	**15**	-	**2544,72**	**21**	*Medium height : 89,75 m*		-

Table 9. Characteristics of DMA A1

1. Integrated devices of remote metering and monitoring FCS (Fluid Control System);

2. Integrated devices of remote metering FMS (Fluid Meter System).

In consideration of the high-tech components and the above-mentioned features (technologies to meter water parameters, feeding modalities, data input and output), FCS is unique, as shown by Patent RM 96 000732 dated 28/10/1996.

All devices are equipped with a data acquisition unit, transmission system and firmware with a license starting at the date of system installation and operation. The FCS allows pressure downstream of the valve to be remotely adjusted, according to specific algorithms supplied by the customer; flow metering is performed by an electromagnetic meter for the 350mm diameter device, whereas for other devices a Woltmann meter with Reed attachments was used; pressure was measured by a pressure transducer.

District extension	**8,87 ha**		
Building type	Prefabricated		
Photos	PHOTO 5 	PHOTO 6 	PHOTO 7 
Buildings	8	8	8
Floors	4	3	5
Apartment for floor	5	6	8
Residential floors	-	1	
Commercial floors	1	-	1
Stores that need water	8	-	4
Public buildings	municipality, 1 elementary school		
Users for type	128	96	260
Users	544		
Enel users	539		
Inhabitants	1937		

Hydraulic parameters	*Pipes*	*Diameter*	*Length [m]*	*Nodes*	*Height max [m]*	*Height min [m]*	*Roughness*
	4	150	850,10	11	105,60	70,00	-
	3	100	550,80				-
TOT	**15**	-	**1400,90**	**11**	*Medium height: 85,83 m*		-

Table 10. Characteristics of DMA A2

In remote metering and remote monitoring devices, pressure is adjusted by a hydrovalve. Devices are fed by an autonomous battery, separate from the connection to the power supply network.

Each remote metering and remote monitoring station is equipped with a system of data acquisition managed by an industrial Microcontroller with acquisition of analogue and digital I/O signals and serial management RS-232. Data are communicated to the control centre via GPRS and the Internet.

The procedure of data communication is such that all stations periodically send data to the Web server, checking the update of their own configuration.

The creation of district in the water supply network of Monterusciello can be achieved by installing the following equipment:

<table>
<tr><td>District extension</td><td colspan="7">9,23 ha</td></tr>
<tr><td>Building type</td><td colspan="7">Prefabricated</td></tr>
<tr><td>Photos</td><td colspan="3">PHOTO 8
</td><td colspan="4">PHOTO 9
</td></tr>
<tr><td>Buildings</td><td colspan="3">4</td><td colspan="4">12</td></tr>
<tr><td>Floors</td><td colspan="3">4</td><td colspan="4">4</td></tr>
<tr><td>Apartment for floor</td><td colspan="3">8</td><td colspan="4">5</td></tr>
<tr><td>Residential floors</td><td colspan="3">▪</td><td colspan="4">1</td></tr>
<tr><td>Commercial floors</td><td colspan="3">▪</td><td colspan="4">▪</td></tr>
<tr><td>Stores that need water</td><td colspan="3">▪</td><td colspan="4">▪</td></tr>
<tr><td>Public buildings</td><td colspan="7">1 high school</td></tr>
<tr><td>Users for type</td><td colspan="3">128</td><td colspan="4">180</td></tr>
<tr><td>Users</td><td colspan="7">308</td></tr>
<tr><td>Enel users</td><td colspan="7">325</td></tr>
<tr><td>Inhabitants</td><td colspan="7">1096</td></tr>
<tr><td rowspan="2">Hydraulic parameters</td><td>Pipes</td><td>Diameter</td><td>Length [m]</td><td>Nodes</td><td>Height max [m]</td><td>Height min [m]</td><td>Roughness</td></tr>
<tr><td>6</td><td>200</td><td>921,10</td><td>7</td><td>68,80</td><td>52,10</td><td>▪</td></tr>
<tr><td>TOT</td><td>6</td><td>▪</td><td>921,10</td><td>7</td><td colspan="2">Medium height: 62,30 m</td><td>▪</td></tr>
</table>

Table 11. Characteristics of DMA B1

- No. 3 integrated FCS devices (150 mm diameter);
- No. 2 integrated FMS devices (150 mm diameter);
- No. 2 integrated FMS devices (200 mm diameter);
- No. 1 integrated FMS device (300 mm diameter);
- No. 1 integrated FMS device (350 mm diameter);

Figures 12-15 show some of the stages of device installation. The manholes, where the experimental devices will be located, were inspected and some of them underwent structural adjustments.

4.5 Remote Control System

The devices selected for the districts are integrated into an advanced Remote Control System (RCS) accessible from the Internet in ASP (Application Service Provider) modality and linked to Google Map for the geo-reference

District extension	**18,37 ha**		
Building type	Prefabricated		
Photos	PHOTO 10 	PHOTO 11 	PHOTO 12 
Buildings	29	10	11
Floors	3	5	2
Apartment for floor	4	6	6
Residential floors	-		-
Commercial floors	-	1	-
Stores that need water	-	6	-
Public buildings	3 school: 1 elementary school, 1 high school, 1 technical school		
Users for type	348	246	136
Users	730		
Enel users	724		
Inhabitants	2599		

	Pipes	*Diameter*	*Length [m]*	*Nodes*	*Height max [m]*	*Height min [m]*	*Roughness*
Hydraulic parameters	3	200	256,64				-
	4	150	494,10	22	59,30	30,00	-
	21	100	2845,80				-
TOT	**15**	-	**3596,54**	**22**	*Medium height: 43,25 m*		-

Table 12. Characteristics of DMA B2

of equipment and visualization of the water supply network. Each FMS and FCS device for metering and monitoring is equipped with a system of data acquisition managed by an industrial Microcontroller with analogue and digital I/O signal acquisition and serial management RS-232 to collect data from sensors, with an internal Compact Flash memory for data and alarms.

Data communication towards RCS centre is carried out via GMS/GPRS and the Internet, whereas the management of alarms is made via SMS, with the possibility to receive asynchronous commands by SMS.

The main advantages of managing a Web centre are:

- accessibility from any workstation connected to the Internet, including new generation hand-phones and mobile phones;

District extension	**27,68 ha**			
Building type	Prefabricated			
Photos	PHOTO 13 		PHOTO 14 	PHOTO 15 
Buildings	3	9	8	15
Floors	5		5	4
Apartment for floor	8	12	6	3
Residential floors	1	1		-
Commercial floors	-	-	1	-
Stores that need water	-	-	8	-
Public buildings	1high school			
Users for type	528		264	180
Users	908			
Enel users	909			
Inhabitants	3232			

	Pipes	*Diameter*	*Length [m]*	*Nodes*	*Height max [m]*	*Height min [m]*	*Roughness*
Hydraulic parameters	1	300	73,85	13	66,00	24,00	-
	5	200	542,05				-
	12	100	2460,72				-
TOT	**15**		**3076,72**	**13**	*Medium height: 46,23 m*		-

Table 13. Characteristics of DMA B3

- possibility of data validation by operators located in several offices;
- data files on a relational database resident on the Customer's server or on a remote server;
- no resident installation required for Client computers;
- easy remote maintenance of the application.

Furthermore, the software is able to recognize several categories of users, each of which is enabled to carry out some operations on monitoring data, in particular:

- USER: it allows validated and non-validated data on specific stations to be examinated;

District extension	**34,37 ha**			
Building type	Prefabricated			Traditional structure
Photos	PHOTO 16 	PHOTO 17 	PHOTO 18 	PHOTO 19 
Buildings	12	13	11	25
Floors	4	4	4	
Apartment for floor	8	8	4	
Residential floors	-	-	-	-
Commercial floors	1	1	-	-
Stores that need water	5	6	-	-
Public buildings	1 curch, 1 school, 1 police building, 1 firman building, 1 Enel, 1 swimming pool			
Users for type	293	318	176	207
Users	994			
Enel users	1022			
Inhabitants	3539			

	Pipes	*Diameter*	*Length [m]*	*Nodes*	*Height max [m]*	*Height min [m]*	*Roughness*
Hydraulic parameters	1	350	146,20	31	74,00	44,80	-
	3	200	481,65				-
	14	150	1123,17				-
	17	100	1889,52				-
TOT	**15**	-	**3640,54**	**31**	*Medium height: 69,20 m*		-

Table 14. Characteristics of DMA C

- SUPERVISOR: it allows all data to be viewed and data relating to a predefined subgroup of stations to be invalidated;
- ADMINISTRATOR: it can access all system functions;
- PUBLIC (without password): it can access only validated data on all stations.

The main functions of the software used for management are shown below:

- **Log in** (Figure 16): it gives access to remote control functions by login and password.

- **Monitor** (Figure 17): it gives the possibility to view the "latest" data sent by devices; furthermore, this function allows the current situation of the system to be monitored and, in case of alarm due to overflow, the relevant data is shown in red:

District extension	**15,4 ha**			
Building type	Prefabricated			
	PHOTO 20	PHOTO 21	PHOTO 22	
Photos				
Buildings	6	8	4	8
Floors	5	3	4	3
Apartment for floor	4	4	4	4
Residential floors	-	-	-	-
Commercial floors	-	-	-	-
Stores that need water	-	-	-	-
Public buildings	1 elementary school			
Users for type	120	144	160	
Users	424			
Enel users	443			
Inhabitants	1509			

	Pipes	*Diameter*	*Length [m]*	*Nodes*	*Height max [m]*	*Height min [m]*	*Roughness*
Hydraulic parameters	1	200	185,90	31	106,40	71,8	-
	15	150	3774,22				-
TOT	**16**	-	**3960,12**	**31**	*Medium height : 89,01 m*		-

Table 15. Characteristics of DMA D

- **Map** (Figure 18): it gives the possibility to view the "latest" data sent by devices; furthermore, this function allows monitoring of the current situation of the system and, in case of alarm due to overflow, the relevant data is shown in red.

- **Data** (Figure 19): it gives the possibility to view the "latest" data acquired by devices; this function also allows monitoring of the current situation of the system and, in case of alarm due to overflow, the relevant piece of information is shown in red.

- **Track** (Figure 20): it gives the possibility to view the "latest" data acquired by devices; this function also allows monitoring of the current situation of the system and, in case of alarm due to overflow, the relevant data is shown in red.

The RCS allowed all available information to be monitored through the customized web interface.

Figure 12. Stage of installation of devices a)

Figure 13. Stage of installation of devices b)

Figure 14. Stage of installation of devices c)

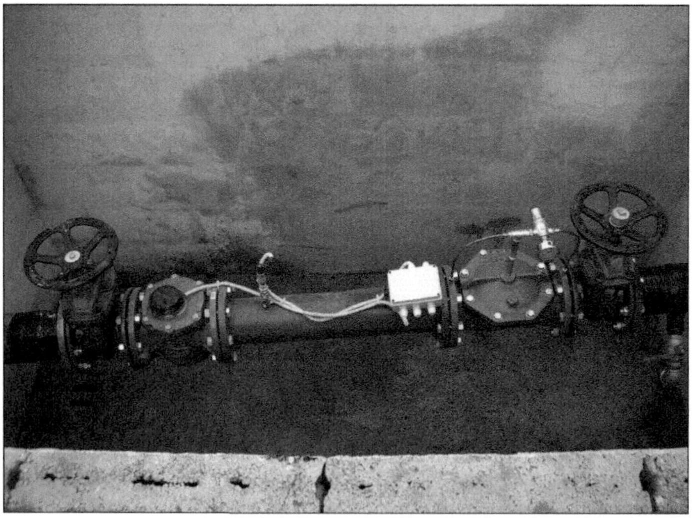

Figure 15. Stage of installation of devices d)

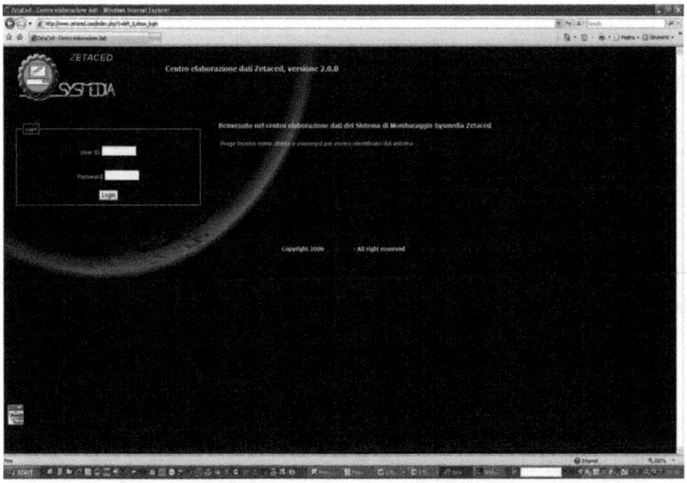

Figure 16. An example of Log in of RCS

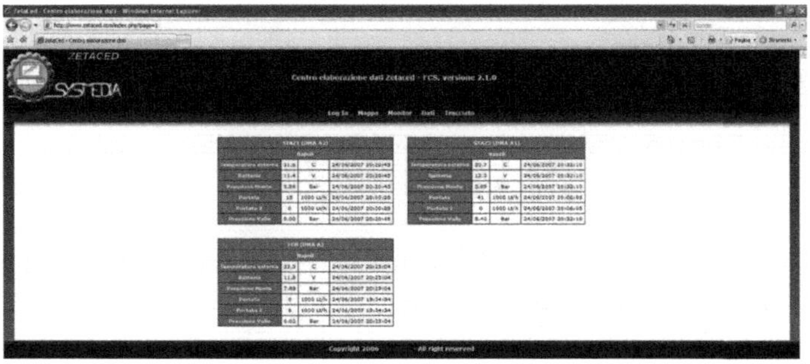

Figure 17. An example of a Monitor of RCS

Figure 18. An example of a Map of RCS

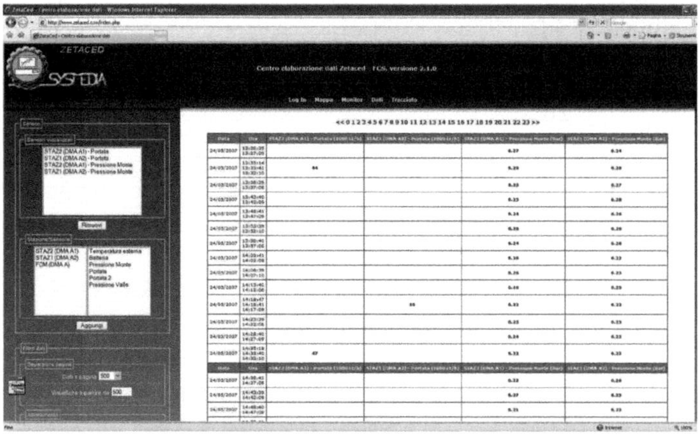

Figure 19. An example of Data function of RCS

In particular, flows and pressures for each installed FCS and FMS have been measured and recorded, as shown in the table below (Table 17), which gives an example of pressure and flow measures between 2:56 pm and 3:54 pm on a "working day" of September 2007 at the stations FCS A1, FCS A2 and FCM A, located in DMA A1 and A2 as illustrated in Figure 21.

Measurements refer to the time when they are recorded and can be aligned by defining a proper time step. Each station is equipped with a timer that, at fixed intervals, synchronizes with the RCS and ensures the synchrony of measurements. The RCS also issues diagrams of recorded measures vs. time, as reported in the following figures (Figure 22-23) showing system information (flows and pressures) at given intervals.

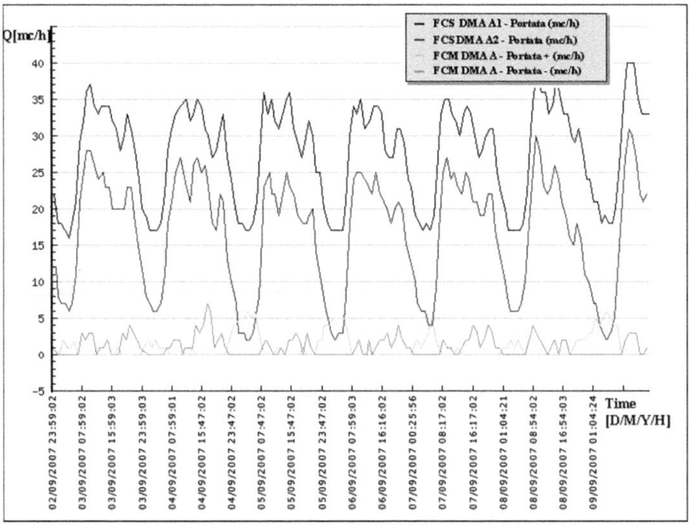

Figure 20. An Example of Track function of RCS

The measuring campaign is still ongoing to improve the calibration of the network model, to calculate consumption at different hours and detect any abnormalities (leakages, low pressures, etc.).

Furthermore, pressure management is being tested in the district DMA A, in order to verify the possibility of decreasing leakages by reducing operation pressures within the limits set by national regulations and suggested by the international literature.

4.6 DMA Pressure Management

Before performing the optimization process for PRV setting, presented in section 3.2, some analyses were made on the data acquired by the remote monitoring system in DMA A1 and DMA A2, as reported in the network model of Figure 21.

To characterise a representative operating condition of the network, multiple time-series of flow and pressure acquired during September 2007 at the three measuring stations (FCS A1, A2 and FMS A3) were averaged excluding holidays.

Preliminary data analysis showed that "working day" and "holidays" were characterised by systematic different behaviours. Working days were then extracted from the original data set, and averaged in order to define

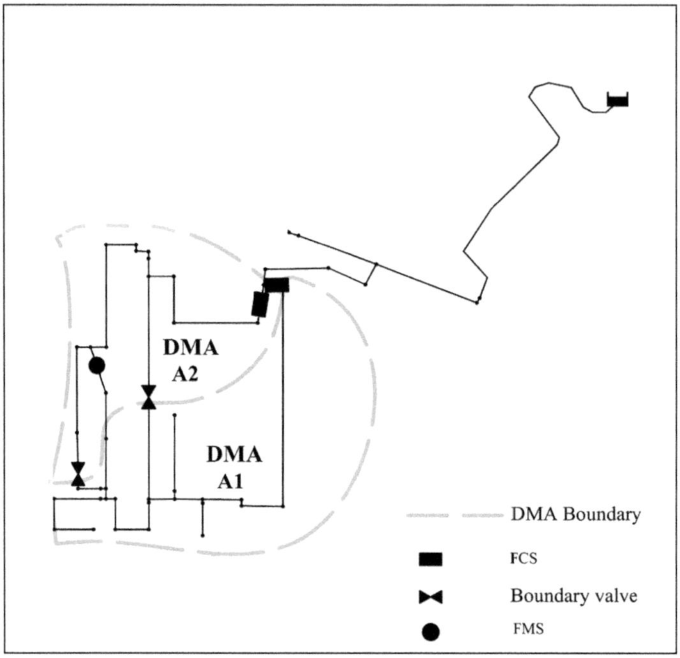

Figure 21. FCS and FMS location of Monterusciello network DMA A1 and A2

the "mean working day".

This data treatment allowed us to reduce the influence of day-to-day variations in hourly demand and to measure uncertainties on the experimental data used to identify the unknown parameters.

September flow and pressure of FCS A1, FCS A2 and FMS A3 were reported in Table 16 and Table 17, respectively. While the September total inflow of DMA A was reported in Figure 24. In Figure 25 "mean working day" flow vs. pressure is reported for FCS A1 and A2, and in Figure 28 the same information is given for FCS A3.

Data analysis of Figures 25 and 26 show different flow and pressure levels in all measurement equipments. Specifically, stations FCS A1 and FCS A2 measures a pressure deviation up to 0.6 bar approximately; while FMS A3 measures a pressure higher than others FCS up to about 1.0 bar because station FMS A3 has a lower elevation. Then Figure 26 shows that FMS A is a bidirectional flow meter with positive and negative flow values.

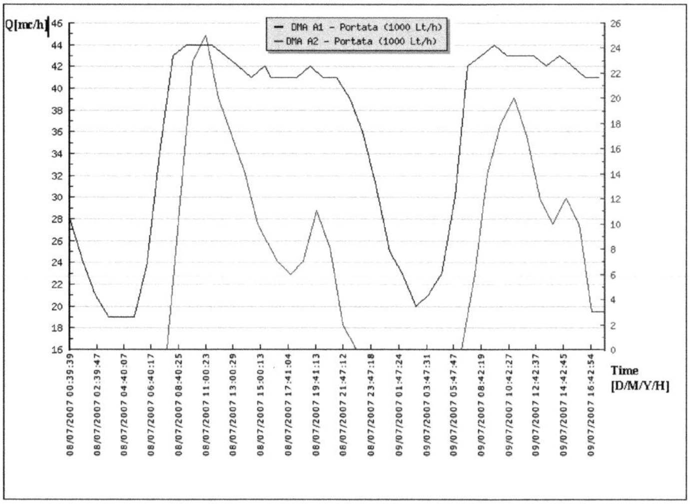

Figure 22. Examples of measurements recorded by RCS (Flow vs. time)

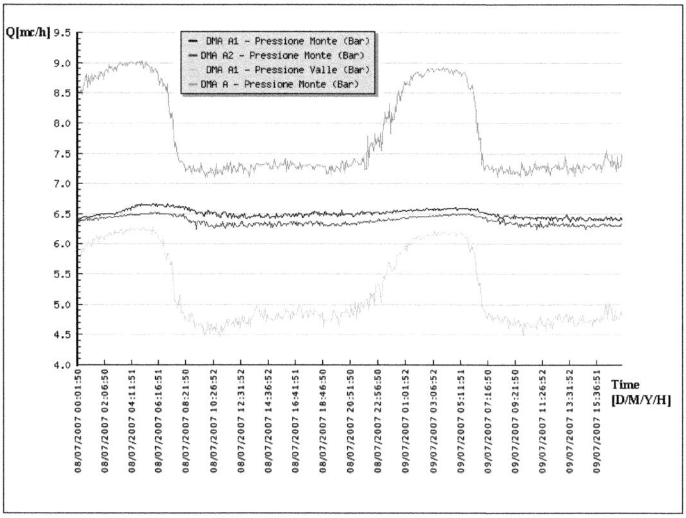

Figure 23. Examples of measurements recorded by RCS (Pressures vs. time)

TIME	FCS A1	FCS A2	FCS A1+A2	MEAN
h	m³/h	m³/h	m³/h	m³/h
0.00	21.13	8.75	29.88	
1.00	18.60	6.25	24.85	
2.00	17.80	5.10	22.90	
3.00	17.55	4.40	21.95	
4.00	17.80	4.25	22.05	
5.00	18.75	4.95	23.70	
6.00	23.60	8.20	31.80	
7.00	32.60	18.60	51.20	
8.00	35.85	26.10	61.95	
9.00	34.90	26.15	61.05	
10.00	34.25	26.20	60.45	
11.00	32.60	24.35	56.95	
12.00	32.23	22.35	54.57	**45.32**
13.00	32.65	22.50	55.15	
14.00	33.10	25.50	58.60	
15.00	33.75	24.92	58.67	
16.00	33.25	21.95	55.20	
17.00	29.55	20.10	49.65	
18.00	29.00	18.85	47.85	
19.00	30.15	19.25	49.40	
20.00	31.95	21.60	53.55	
21.00	31.85	21.35	53.20	
22.00	28.40	16.50	44.90	
23.00	25.55	12.70	38.25	

Table 16. Mean of the hourly flow during September "working day" of FCS A1 and A2

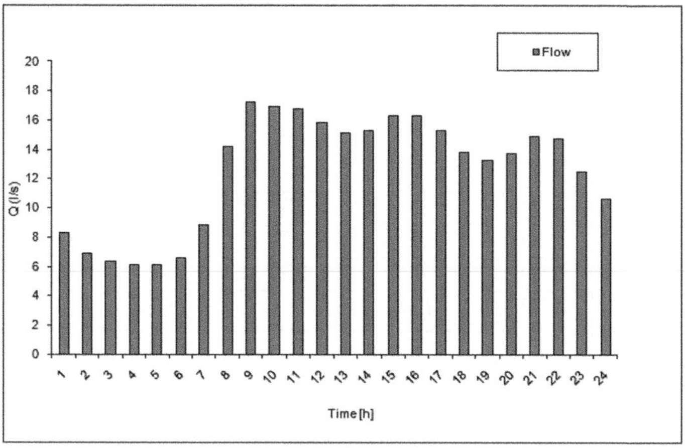

Figure 24. Mean of the hourly flow during September "working day" of FCM A

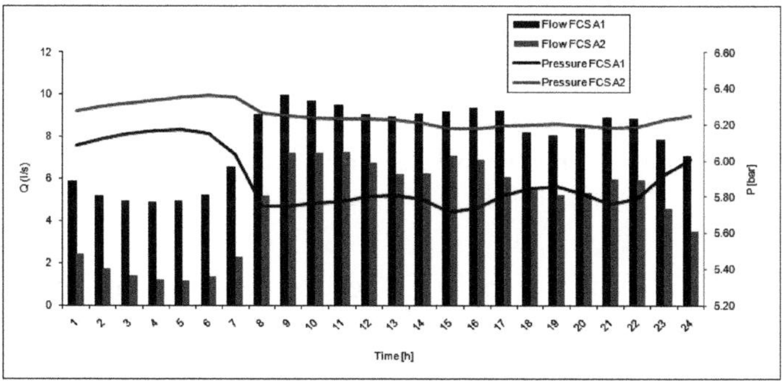

Figure 25. September "mean working day" flow in DMA A

Time	FCS A1	FCS A2	FCS A
h	bar	bar	bar
0.00	6.08	6.28	7.50
1.00	6.13	6.30	7.52
2.00	6.15	6.32	7.53
3.00	6.16	6.34	7.53
4.00	6.17	6.35	7.52
5.00	6.15	6.37	7.50
6.00	6.03	6.35	7.44
7.00	5.75	6.27	7.29
8.00	5.75	6.25	7.26
9.00	5.77	6.24	7.26
10.00	5.78	6.23	7.28
11.00	5.80	6.23	7.31
12.00	5.81	6.23	7.32
13.00	5.79	6.21	7.32
14.00	5.72	6.18	7.23
15.00	5.74	6.18	7.28
16.00	5.81	6.19	7.33
17.00	5.84	6.20	7.34
18.00	5.86	6.20	7.38
19.00	5.81	6.20	7.34
20.00	5.76	6.18	7.30
21.00	5.79	6.19	7.34
22.00	5.92	6.22	7.43
23.00	6.01	6.25	7.47

Table 17. September "mean working day" flow vs. pressure of FCS A1 and FCS A2

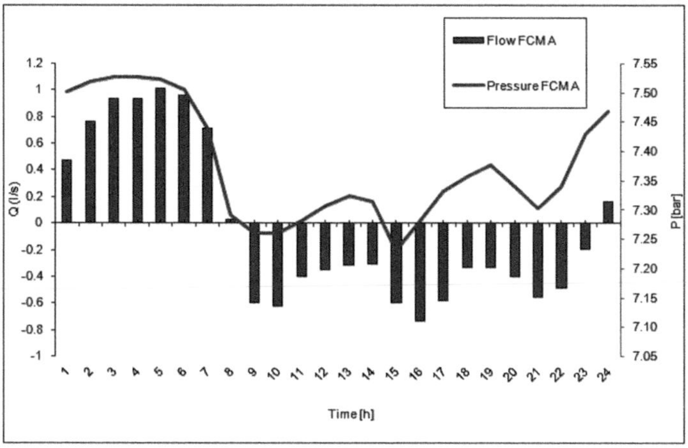

Figure 26. September "mean working day" flow vs. pressure of FMS A

Following the flow chart of network PRV setting, reported in Figure 3, the above figures illustrate the results of the identification procedure (step a) applied to the data set in Tables 16 and 17.

Figure 27 compares measured and computed flow and pressure time series, respectively. As may be observed, the time behaviour of the flow is very well reproduced for both the two measuring stations on the feeder pipes (FCS 1 and 2) and that on the internal pipe (FMS 3).

The three peaks observed in the measured inflow are clearly recognizable in the corresponding simulated hydrographs, and the repartition of the inflow between the two feeding pipes is also well reproduced.

Also the temporal behaviour of pressure (Figure 28) exhibits a reasonable qualitative agreement, although some systematic quantitative differences exists, especially for measuring station 3.

The computed values of the parameters are listed in Table 18: indeed, the estimated per-capita net freshwater daily consumption (about 215 l) and hourly peak coefficient (about 1.6) agree with the values reported in the literature for residential areas with a comparable population (Stephenson, 1998).

The exponent in the leakage law (1) is relatively close to the commonly used value b = 0.5 of an ideal orifice flow.

The values of Table 18 were used to optimize PRV setpoints or similarly the coefficient a_j, as described in section 3.2 (step b).

In particular, in order to better reproduce the behaviour of FCS installed

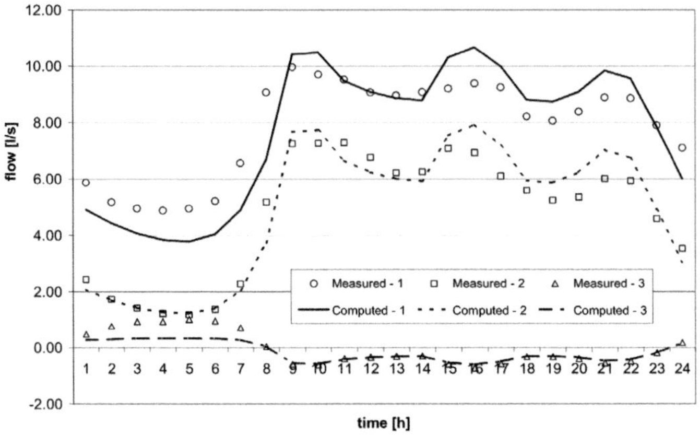

Figure 27. Measured and computed flow time-series [27]

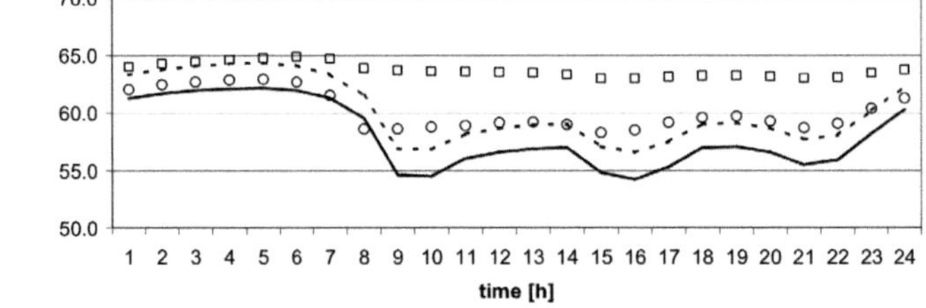

Figure 28. Measured and computed pressure time-series [27]

k	1	2	3	4	5	6	7	8	9	10	11	12
c^d_k (· 1000)	1.01	0.83	0.70	0.63	0.61	0.70	1.00	1.71	3.27	3.30	2.87	2.71

k	13	14	15	16	17	18	19	20	21	22	23	24
c^d_k (· 1000)	2.61	2.58	3.22	3.37	3.08	2.59	2.56	2.71	3.02	2.91	2.18	1.43

a	0.011
b	0.569

Table 18. Estimated values of model parameters [27]

at the pilot site, control valves were modelled as PRV in EPANET 2.0 [75]: therefore, the results of the optimization procedure consist in the optimal adjustment of piezometric altitude of each valve.

Operational pressure levels are influenced by variations in demand. Nevertheless, we may make a distinction of two separate time periods: night-time, when the operational pressure ranges between 6.0 and 6.5 bar, and day time when the flow pressure ranges between 5.5 and 6.0 bar.

According to the analysis of the different operational conditions (peak and night consumption) and the minimum performance levels for users, different pressure adjustments were carried out on FCS A1 and FCS A2 districts by comparing the trend of hourly flows on different week days.

Specifically the valve daily setpoint was then split into two time slices as homogeneous as possible, identified as follows:

- Time slice 1 : h 0:00 - 6:00 am (night flow)
- Time slice 2: h 7:00 am - 23:00 pm (daytime flow)

With reference to each time slice, application of the optimization procedure, described in section 3.2, allowed us to define the optimal adjustment of two network FCSs, whose location is shown in Figure 21. Through a penalization mechanism, it was fixed that the minimum piezometric altitude of the network could not be lower than 20 m.

Pressure adjustments by hydrovalves were made in the whole DMA A district.

In this district the total hourly network inflow in DMA A1 and DMA A2, recorded from May 2007 to September 2007, which is obviously influenced by the season and daily changes, shows an average value of about 42 m^3/h, and peaks above 60 m^3/h during higher consumption hours; average values recorded during night-time are slightly below 20 m^3/h.

In particular, starting from the results from the simulations and from the optimization model developed, reported synthetically in Figure 29, two different pressure thresholds were identified in the first stage, instead of three as stated in the previous section.

Thus the automatic remote monitoring system was adjusted to two different pressure levels, reported in Figure 21

a) Time slice 1: h 0:00 - 6:00 am (night flow): 3.5 bar;
b) Time slice 2: h 6:01 am - 11:59 pm (daytime flow): 5.0-6.0 bar.

Thus the automatic remote monitoring system was adjusted to two different pressure levels: the condition of Time band 1 was initially gradually achieved by decreasing the operation pressure by 0.3 bar/day, in order to avoid creating problems to users.

Figures 30-31-32 below show both the pattern of pressure and that of flows on 12, 13 and 18 October 2007.

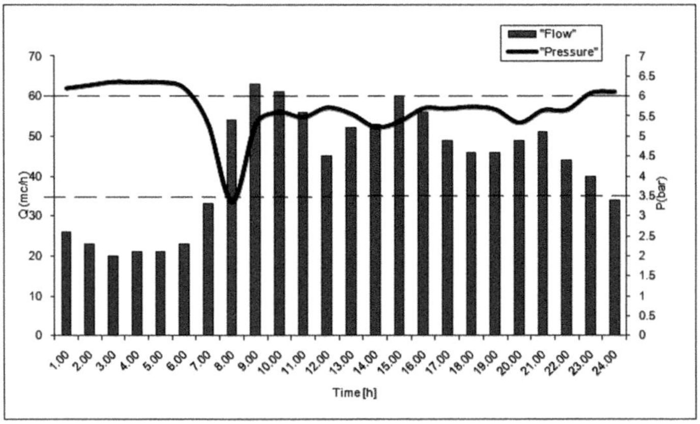

Figure 29. Optimal pressure adjustment

From the analysis of diagrams, which refer to different week days, it seems evident that night-time pressure adjustments allows a considerable reduction in night flows.

For the sake of clarity, Figure 31 shows pressure and flow trends on Thursday 18/10/2007 (with adjustment) compared with unadjusted average or the same week-day during the month. In particular, between 0.00 am and 6.00 am, the control system was bound to keep a pressure value of 3.50

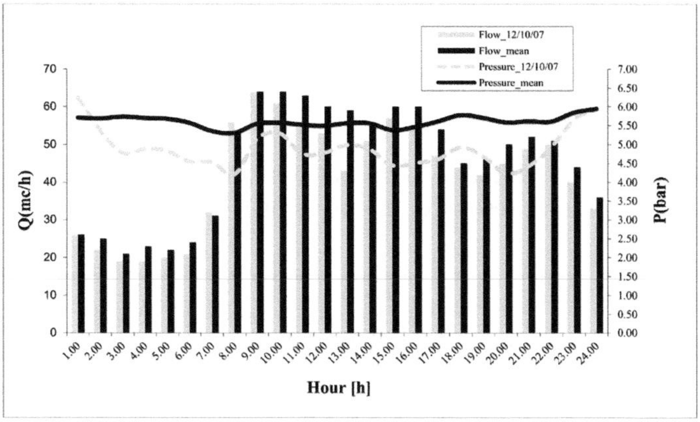

Figure 30. Values of inflows on Friday 12/10/2007 (adjusted) and on other Fridays of the month (unadjusted averages) with relevant pressure trends)

bar ± 0.5 bar.

By doing so, there was a saving of water resources equal to about 15% of the total flow consumed in minimum consumption hours.

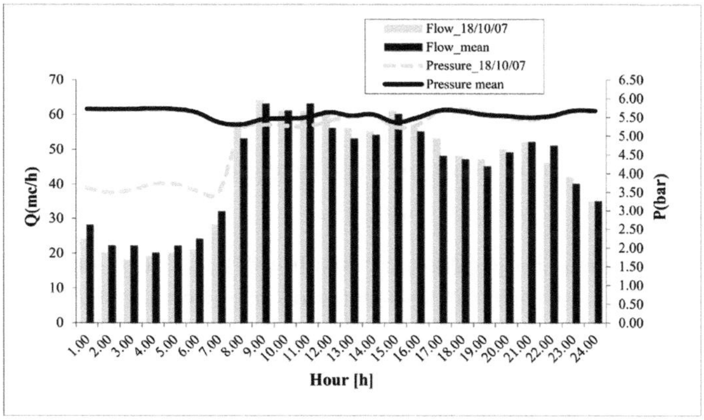

Figure 31. Values of inflows on Thursday 18/10/2007 (adjusted) and on other Thursdays of the month (unadjusted averages) with relevant pressure trends

Finally, Figure 32 reports the pressure and flow trends on "holidays day" Saturday 13/10/2007, (with adjustment) in comparison with the average values on the same week day during the month (without adjustment).

In this case, in particular, adjustment was carried out during night-time, between 3.00 am and 8.00 am, setting the control system at a pressure of 4.50 bar ± 0.5 bar as well as during the day, with a pressure value about 1 bar lower than ordinary operation.

Also in this case, a different choice of pressure managment shows a significant saving of meter resources,in excess of 10% .

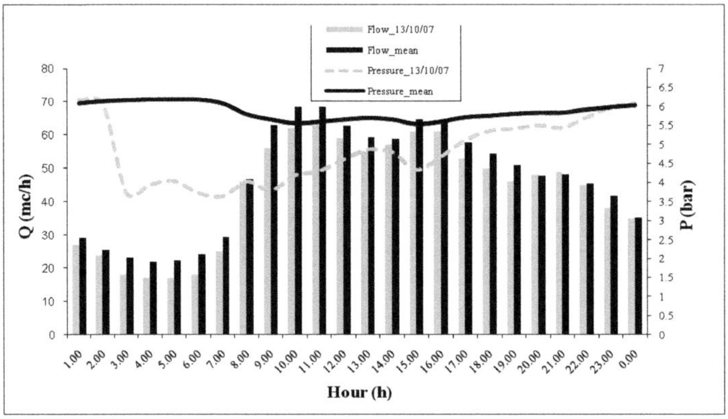

Figure 32. Values of inflows on Saturday 13/10/2007 (adjusted) and on other Saturdays of the month (unadjusted averages) with relevant pressure trends

4.7 Estimation of cost and savings with pressure control

This chapter reports the preliminary results of pressure management combined with district metering in the water distribution network of Monterusciello.

Starting from the estimated consumption in the DMA A1 and DMA A2 districts and from flow and pressure measurements metered by installing FCS and FMS systems, the method proposed in the previous section was implemented and gave interesting results both in terms of prelocation of water losses and resource saving thanks to pressure adjustments at various times of the day.

Using the data in Table 17, the comparison of the Minimum Night Flow district in DMA A1 and A2 with those suggested by the British literature was very useful. As explained below, it allowed night losses to be estimated and the launching of policies to reduce wastes.

DMA A	A1	A2	Tot
Users	2777	1937	4714
Inhabitants	764	539	1303
Pipe length [km]	2.6	1.5	4.1
AZNP [m]	72.3		
Non-household users	1 church, 1 bank, 1 post office	Town Hall, 1 primary school	

Table 19. DMA A1 and A2 data

The main goals of the tests were grouped into several categories:

- evaluation of the district water balances;
- interpretation of the Minimum Night Flow (MNF)
- effects of the differentiated adjustment of district pressures;
- evaluation of the economic advantage of the intervention.

The first flow and pressure measurements and subsequent processing suggest a positive outcome of the results of tests and goals. Then, by comparing results of pressure management with the law on the detection procedure of leakage it is possible to make some considerations.

Based on present results, the Monterusciello network seems to be characterized by a rather constant value of the leakage level, with small variations from one hour to the next (0.8-1.0 l/s).

Leakage amounts to about 7% of the inflow volume (Figure 30), a value consistent with the relatively young age of the hydraulic network.

Nevertheless this leakage percentage, as will became clear below, was significantly underestimated by identification method used in order to define the leakage law (26) and design pressure management setpoints.

The preliminary analysis of night use, compared with the number of users in district A, proved that it is necessary to examine the Minimum Night Flow in detail with the methodology proposed by "Managing Leakage: UK Water Industry Managing Leakage" [104] which provides for the estimate of the following items:

-EXCEPTIONAL CUSTOMER NIGHT USE: any customers using a flow higher than 500 l/hr during the night have to be considered as exceptional; furthermore, they have to be identified separately and added individually to the estimate of minimum night flow;

-ASSESSED NORMAL HOUSEHOLD NIGHT USE: flow used at night by household users; it can be calculated as:1.7 [l/household/hr] * N_H (number of household users) or 0.6 [l/resident/hr] * n (number of residents);

-ASSESSED NORMAL NON-HOUSEHOLD NIGHT USE: flow used at night by users other than household users (schools, police stations). In a simplified way, it is possible to calculate the value of this flow as 78 [l/non-household/hr] * $N_N H$ (number of non-household users). However, it is possible to split the several kinds of non-household users into five groups from A to E, to which a specific flow value is assigned according to the REPORT E of the "Managing Leakage: UK Water Industry Managing Leakage" [104];

-BACKGROUND NIGHT FLOW LOSSES: these represent background losses both in main and secondary pipes. Generally, they depend on the conditions of the infrastructure and on the pressure. Three values are proposed for infrastructures: "good", "average" and "fair" [104], each of them corresponding to an Average Zone Night Pressure (AZNP) equal to 50m. For an AZNP other than 50m, it is necessary to adjust the obtained value through the multiplying coefficient PCF (Pressure Correction Factor).

These items, useful also to determine water balances, are estimated in order to identify the presence of possible undetected bursts. In the case of the Monterusciello DMA A, the data used to calculate the minimum night flow are reported in Table 20; furthermore, the presence of three exceptional household customers and one condition of "average" infrastructure is assumed. Non-household customers were not taken into account since they were not active at night.

The calculation of the minimum night flow for DMA A is reported in Table 20 below.

Analysis of the MNF shows that the total background minimum night flow measured in l/h should be equal to 13,146 l/h, that is 13.146 m^3/h, with a variance of 0.137 m^3/h, thus ranging between 13.009 and 13.316 m^3/h, whereas the average night flow measured is equal to 21.85 m^3/h.

Therefore, the difference amounting to 8.74 m^3/h seems to be quite anomalous and attributable to undetected bursts in the network. The subsequent zooming stage, is a further division of district A into two sub-districts

A1 and A2. Pressure adjustment will give more information on this issue.

The possible presence of pre-existing losses due to bursts can be further supported following the event described in Figure 33, which shows the flow measures of FCS 2 between 12:00 am on 1 June 2007 and 12:00 am on 4 June 2007.

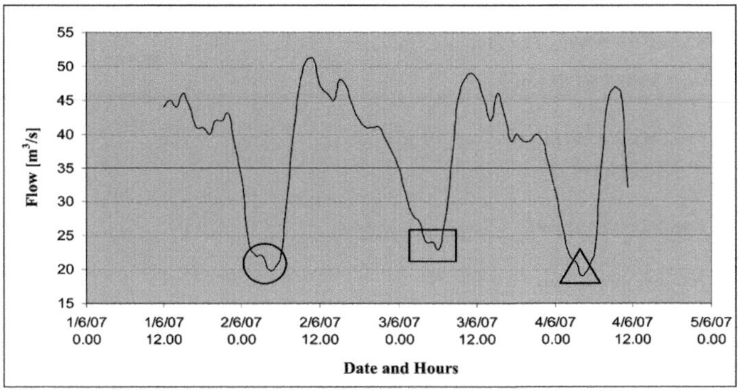

Figure 33. Flow measures FCS 2 (Leakege Detection and Repair)

Indeed, it is possible to see the occurrence of a water leakage promptly detected by the system of continuous network monitoring.

Figure 33 shows the night flow on June 2 2007, in the circle, (the average between 2.00 and 6.00 am is equal to about 21 m^3/h as during the previous days - a high value as previously explained); the following night, June 3 2007, the measuring station signals an increase in the flow with values of 24 m^3/h during the same time interval, in the rectangle.

Finally, after maintenance, the flow decreases to an average 19 m^3/h (triangle) during the night of June 4 2007.

Specifically, during ten experimental days (from 12th October to 18th October 2007) water saving was above 16% during the night and about 7% in the rest of the day, as reported in Figure 34 (total flow) and Figure 35 (night flow) in which four pressure management days are in grey while six undisturbed system days are in black.

These results, which were achieved without complaints from users due to possible inconveniences, confirm that it is appropriate to make differentiated adjustments of pressure in order to reduce overloads which, are known to cause considerable night losses and can cause breaks on already stressed pipes.

A possible saving forecasting, relative to the whole Monterusciello network, based on extrapolation of these experimental results, was then esti-

CUSTOMER NIGTH USE			**COMPONENTS**	**SUBTOTAL l/hr**	**TOT l/hr**
SUM OF EXCEPTIONAL NIGHT USERS		**H/HOLDS**	**1500**	**1500**	
> 500 l/hr individually		**NON H/HOLDS**			
ASSESED HOUSEHOLD NIGHT USE:					
1,7 l/household/hr x No of properties (N_H) or			**2830,20**	**2830,2**	
0,6 l/resident/hr x No of residents (n)		**0,6 x 4717**			
ASSESED NON-HOUSEHOLD NIGHT USE:					
		A			
SIMPLIFIED: **8 l/non-household/hour x**		**B**			
N° of non-household (N_H) or		**C**			
		D			
DETAILED: **Classified by Groups A to E**		**E**			
				=	**4330,20**
BACKGROUND LOSSES AT 50M AZNP					
	CONDITION	L/KM/HOUR	LENGTH (KM)	SUB TOTAL	**TOT km**
DISTRIBUTION MAINS	**Good**	**20**			
	Average	**40**	**4,1**	**164**	
	Fair	**60**			
	CONDITION	L/PROP/HOUR	NO OF PROPS		
	Good	**2,0**			
SERVICES	**Average**	**4,0**	**1303**	**5212**	
	Fair	**6,0**			
					5376

AZNP (metres)	**72,3**	**x PCF**	**1,64**	**8.816,64**
PRESSURE CORRECTION FACTOR (PCF)				

AZNP	**20**	**25**	**30**	**35**	**40**	**45**	**50**	**55**	**60**	**65**	**70**	**75**	**80**	**85**	**90**	**95**	**100**	**105**	**110**	**115**	**120**
PCF	**,33**	**,43**	**,53**	**,64**	**,75**	**,87**	**1,00**	**1,13**	**1,27**	**1,42**	**1,57**	**1,72**	**1,88**	**2,05**	**2,23**	**2,41**	**2,59**	**2,78**	**2,98**	**3,18**	**3,39**

TOTAL BACKGROUND MINIMUM NIGHT FLOW (l/hr)				**13.146,84**
STANDARD DEVIATION OF ON-OFF MEASUREMENT			SUBTOTAL l/hr	**TOTAL l/hr**
ASSESSES HOUSEHOLD NIGHT USE	$3{,}8\sqrt{N_H}$ or $2{,}4\sqrt{n}$	$3{,}8 \times \sqrt{1303}$	**137,17**	**137,17**
ASSESSES NON-HOUSEHOLD NIGHT USE	$11\sqrt{N_{NH}}$			

Table 20. Minimum Night Flow calculated with Water Industry Research method

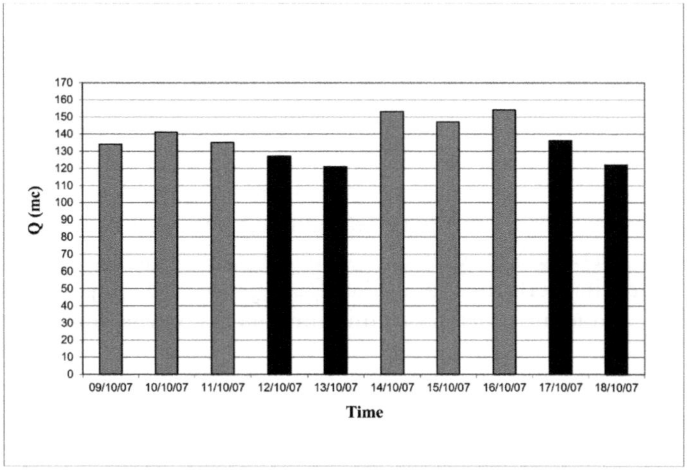

Figure 34. Night Flow

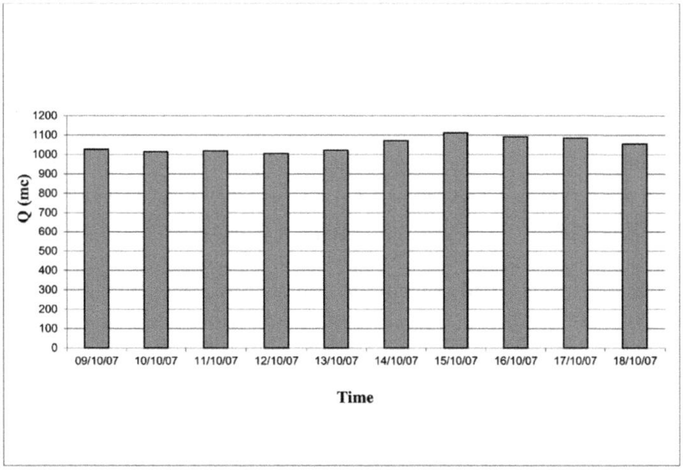

Figure 35. Total Flow

mated with a water saving comprised between 5 and 17%.

4.8 Conclusion

The research confirmed the effectiveness of the two techniques, namely District Metering and Pressure Management, implemented at the Monterusciello pilot site, in order to detect and decrease water leakage.

In particular, Design Support Methodology allowed us to define permanent Water District Metering of a water supply network compatible with system hydraulic performances. The DSM, based on graph theory principles and on the identification of minimum energy paths, allowed us first to define the Main Network Layout and then to draw the Main Graph G_M with specific features to simplify WDM design helping operators.

The results, obtained by testing DSM on the Monterusciello network, showed that DSM rapidly identifies WDMs with satisfactory performance indices and allows all design constraints to be met. The DSM proposed is robust because, even for large networks that have many possible WDMs, DSM allows rapid identification of good DMA layouts with resilience index values close to original network layout. Furthermore DSM is very flexible too because it provides WDM planners with some alternative DMA layouts with different numbers of flow meters and operational costs.

DSM may provide a valid technical support to water utilities in Water District Metering design overcoming empirical approaches and, at the same time, respecting water system constraints. The DSM proposed is a flexible tool applicable to networks of any size using commercially available software (e.g. EPANET and MATLAB).

The preliminary results obtained on the pilot site with the differentiated pressure adjustment, districts in A1 and A2, proved the effectiveness of the pressure management methodology to save water resources especially during the night.

The automatic control system of single devices-which is performed via SMS and, in ASP modality, via the Internet - allowed various pressure levels to be fixedin single districts and at different hours of the day, so that the manager can operate with great flexibility.

This possibility of acting remotely on complex water distribution networks gives a real opportunity to make several configurations of the network (temporary or permanent district metering, differentiated pressure adjustment, definition of pre-alarm status, etc.), preliminarily anticipated by simulation analysis, which correspond to the various system scenarios.

Furthermore, the first experiences on the site created by CIRIAM are positive for the efficiency and reliability of the devices chosen for remote controlling. In addiction, the Remote Control System is easy and fast to use and it can be looked up in the web via the Internet.

Our research proved that the development of new technologies, especially in the field of metering and remote controlling, combined with higher efficiency and lower costs, can support their use in all situations where the waste of water resources cannot be accepted in a modern public service.

The quantity of available data is such that they can give the manager a set of important information to manage the water distribution system efficiently, economically and effectively way, which is indeed the aim of existing Italian regulations.

Finally, the pilot site of Monterusciello may be an important opportunity for the Italian technical and scientific community to study the various issues related to the optimization of the management of water distribution systems, by allowing testing of techniques, method and technological innovations developed in Italy and abroad.

Bibliography

[1] AWWA (American Water Works Association). *Manual of Water Supply Practices M36: Water audits and leak detection*, 1997.

[2] AWWA (American Water Works Association). *Water audits and leak detection. 2nd ed.* AWWA manual M36, Denver, 1999.

[3] Alegre, H., Hirner, W., Baptista, J., Parena, R., *Performance Indicators for Water Supply Services*, IWA Manuals of Best Practice, Hardback, 2000.

[4] Alonso, J.M., Alvarruiz, F., Guerriero, D., Hernandez, V., Ruiz, P.A., Vidal, A.M., Martinez, F., Vercher, J., Ulanicki, B., Parallel *Computing in Water Network Analisys and Leakage Minimization*, J. of Water Planning and Management, ASCE (126)4, pag.251-260, 2000.

[5] ARPA-EmiliaRomagna, *Studio sulla ricerca delle perdite degli acquedotti regionali, sulle metodologie di raccolta ed elaborazione dei dati ai fini della loro confrontabilit, sull'analisi dei livelli di perdita economicamente e ambientalmente accettabili, per la definizione di linee guida regionali*, Regione Emilia Romagna, 2005.

[6] Artina, S., Bocchi, C., Bragalli, C., Zorzi, Y., *Approccio metodologico alla distrettualizzazione delle reti di distribuzione*, 28 Convegno di Idraulica e Costruzioni Idrauliche, Potenza, 2002.

[7] Berge, C. *Thorie des graphes et ses applications*, Collection Universitaire de Mathmatiques. Paris: II Dunod (English edition, Wiley 1961), 1958.

[8] Biggs, N., Lloyd, E. and Wilson, R. *Graph Theory* 1736-1936. New York: Oxford University Press., 1986.

[9] Bondy, J.A. and Murty, U.S.R. *Graph Theory with Applications.* New York: Elsevier North-Holland, 1976.

[10] Butler, D. *Leakage detection and management*,Palmer Environmental, UK, 2000.

[11] Calza, F., Delsodato, G., *Esperienza di AGAC sulla distrettualizzazione dell'acquedotto di Reggio emilia: metodi, risultati, valutazioni, La distrettualizzazione delle reti di distribuzione idrica nella misura e nella riduzione delle perdite*, collana Federgasacqua, Bologna 2002.

[12] Caputo, A.C. *An inverse approach for piping networks monitoring*, Journal of Loss Prevention, 2002.

[13] Carpano, M.J. *Automatic Display of Hyerarchized Graphs fo Computer Aided Decision Analysis.* IEEE Trans. Syst. Man Cybern., SMC-10 (11), 705-715,1980.

[14] Cascetta, F., *Sistemi di telecontrollo di reti di pubblica utilità: Reti idriche*, Franco Angeli, Milano, 2003.

[15] Cascetta, F., Di Nardo, A., Di Natale, M., Iervolino, M. and Greco, R. *Analisi degli effetti della distrettualizzazione sull'efficienza di una rete di distribuzione idrica.* EIDOS - La Rivista del Metering, 4, 48-56, 2005a.

[16] Cascetta, F., Di Nardo, A., Di Natale, M., *La distrettualizzazione delle reti idriche di distribuzione*, Acqua&Aria, N.3 Marzo, 2005b.

[17] Cascetta, F., Di Nardo, A., Di Natale, M., *Distrettualizzazione fisica e virtuale delle reti idriche*, Acqua&Aria, N.5 Maggio, 2005c.

[18] Cascetta, F., Di Nardo, A., Di Natale, M., Gisonni, C., and Greco, R., *Una metodologia per la distrettualizzazione di una rete idrica di distribuzione: il caso di Villaricca (NA), La ricerca delle perdite e la gestione delle reti di acquedotto.* Secondo Seminario, Perugia, Morlacchi Editore, Milano, 2006.

[19] Charalambous, B., *Experiences in DMA redesign at the water board of Lemesos,* Cyprus, [online]. Proceedings of Leakage 2005, 12-14 September 2005 Halifax. Available from: http://waterloss2007.com/Leakage2005.com/unterseiten/final program.htm [Accessed 28 February 2009], 2005.

[20] Covas, D., Jacob, A., Ramos, H. *Bottom-Up Analysis for Assessing Water Losses: A Case Study*, Proc. of the 8th Annual Water Distribution Systems Analysis Symposium 2006, Cincinnati, Ohio, USA, 2006.

[21] D'Anisi, L., Lanave, V., *Telecontrollo dei sistemi idrici complessi: la gestione ottimizzata della distribuzione idrica nell'abitato di Potenza,* Atti del I Convegno di Idraulica Urbana, Sant'Agnello (NA), 53-55, sett. 2005.

[22] Deuerlein, J.W. *Decomposition Model of a General Water Supply Network Graph.* Journal of Hydraulic Engineering, 134 (6), 822-832, 2008.

[23] Di Battista, G., Eades, P., Tamassia, R. and Tollis, I.G. *Graph Drawing (Algorithms for the visualization of grafh).* New Jersey (USA): Prentice Hall, 1999.

[24] Di Nardo, A., Di Natale, M., Greco, R., Iervolino, M., *Controllo differenziato delle pressioni per la riduzione delle perdite in una rete idrica distrettualizzata,* Atti del XXX Convegno di Idraulica e Costruzioni Idrauliche, Roma, 2006; pp. 1-11; L322, Editrice La Sapienza, Roma, 2006.

[25] Di Nardo, A., Di Natale, M. and Greco, R. *Valutazione dei costi di distrettualizzazione di una rete idrica ai sensi del D.M. n99 del 1997 nel rispetto delle prestazioni idrauliche di progetto,* In: B. Brunone, M. Ferrante and S. Meniconi, eds. La ricerca delle perdite e la gestione delle reti di acquedotto, 20-21 Settembre 2007 Perugia. Perugia: Morlacchi Editore: 327-334, 2008.

[26] Deuerlein, J.W., 2008. Decomposition Model of a General Water Supply Network Graph. Journal of Hydraulic Engineering, 134 (6), 822-832.

[27] Di Nardo, A., Di Natale, M., Iervolino, M., *Demand Pattern And Leakage Identification In Water Distribution Networks, Proceedings of 33rd* IAHR 2009 Congress - Water Engineering for a Sustainable Environment, Vancouver, pp.1545-1552, 2009.

[28] Dijkstra, E.W. *A note on two problems in connexion with graphs*, Numerische Mathematik, 1, 269-271, 1959.

[29] Fantozzi, M., Lambert, A., *Experiences in application of IWA approach to small water networks in Italy*, H2O Workshop Water Losses Management, Ferrara, maggio 2004.

[30] Farley, M., Trow, S. *Losses in water distribution networks*, Iwa Publishing, London, 2007.

[31] Ferrante, M., Brunone, B., *Pipe system diagnosis and leak detection by unsteady-state tests. 2. Wavelet analysis*, Advances in Water Resources 26 (1): 107-116, 2003.

[32] Germanopoulus, G., Jowitt, P.W. *Leakage Reduction by excessive pressure minimization in a water supply network*, Proc. Inst. Of Civil Engr., 87(2), 195-214, 1989.

[33] Giugni, M., Romanelli, D., Fontana, N., Corrado, V., *Primo approccio alla distrettualizzazione idrica di Napoli*, Atti del I Convegno di Idraulica Urbana, Sant'Agnello (NA), 63-65, 2005.

[34] Giustolisi, O., Kapelan, Z. and Savic, D.A. *An algorithm for automatic detection of topological changes in water distribution networks*, ASCE Journal of Hydraulic Engineering, 134 (4), 435-446, 2008.

[35] Goldberg, D.E. *Genetic Algorithms inSearch, Optimization, and Machine Learning*, Addison-Wesley, 1989.

[36] Goulter, I.C., Morgan, D.R. *An integrated approach to the layout and design of water distribution networks.* Civ. Engrg. Sys., 2 (2), 104-113, 1985.

[37] Greco, M., Di Cristo, C. *Calibration for hydraulic network simulation*, Atti del convegno Water Resources Engineering Conference, Seattle, 1999.

[38] Greyvenstein, B., Van Zyl, J.E. *An Experimental Investigation into the Pressure Leakage Relationship of some Failed Water Pipes.* Water Institute of Southern Africa Biennial Conference, Durban, South Africa, 22-24 May 2006, 2006.

[39] Huang, Y.C. *Enhanced genetic algorithm-based fuzzy multi-objective approach to distribution network reconfiguration.* IEEE Proc. Gener. Transm. Distrib., Vol. 149(5), 615-620, 2002.

[40] Available from http://www.waterportal.com

[41] Jacobs, P. and Goulter, I.C., 1989. Optimization of redundancy in water distribution networks using graph theoretic principles. Eng. Opt., 15, 71-82.

[42] Johnson, D.B. *Efficient algorithms for shortest paths in sparse networks.* Journal of the ACM, 24 (1), 1-13, 1977.

[43] Jowitt, P.W., Xu, C. *Optimal Valve Control inWater Distribution Networks,* J.Water Resour. Planning Manage., ASCE 116(4), 455-472, 1990.

[44] Kapela, Z., Covas, D. *Inverse Transiet Analysis in pipe Networks for leakage detection,* XXX IAHR Congress, Thessaloniki, Greece, 2002.

[45] Kapelan, Z.S., Savic, D., Walters, G.A. *A hybrid inverse transient model for leakage detection and roughness calibration in pipe networks,* Journal of Hydraulic Research, Vol. 41, No. 5, pp. 481-492, 2003.

[46] Kesavan, H.K., Chandrashekar, M. *Graph-theoretic models for pipe network analysis.* Journal of the Hydraulics Division, 98 (HY2), 345-363, 1972.

[47] Lalonde, A. *Flow Modulated Pressure Management in York Region, Ontario - Canada.* [online]. Proceedings of Leakage 2005, 12-14 September 2005 Halifax. Available from: http://waterloss2007.com/Leakage2005.com/unterseiten/final program.htm [Accessed 28 February 2009], 2005.

[48] Lamanna, N.V., Andreazza, C., Cotzias, A., *Piano di valutazione delle perdite nelle reti di distribuzione idrica, pianificazione degli interventi necessari e delle attivit di controllo e monitoraggio, in La distrettualizzazione delle reti di distribuzione idrica nella misura e nella riduzione delle perdite,* collana Federgasacqua, Bologna 2002.

[49] Lambert A., Hirner, W. *Losses from water supply systems. Standard terminology and recommended performance measures.* IWA Blue Pages, International Water Association.
Available from: http://www.iwahq.org/, 2000.

[50] Lambert, A.O., and Hirner, W. *Losses from water supply systems. Standard terminology and recommended performance measures [online].* IWA Blue Pages, International Water Association.
Available from: http://www.iwahq.org/ [Accessed 27 February 2009], 2000.

[51] Leone, G., *Il risparmio idrico attraverso la riduzione delle pressioni di rete, Hydrogeo,* Rimini, 2001.

[52] Levin, B. *Pressure Management in the Pittsburgh Area: A Working and Economical Solution.* [online]. Proceedings of Leakage 2005, 12-14 September 2005 Halifax. Available from: http://waterloss2007.com/Leakage2005.com/unterseiten/final program.htm [Accessed 28 February 2009], 2005.

[53] Liberatore, S., Sechi, G.M. *Ottimizzazione dell'uso di valvole per il controllo in tempo reale della pressione nelle reti acquedottistiche,* In: M. Franchini and P. Bertola, eds. Approvvigionamento e Distribuzione Idrica: Esperienza, Ricerca ed Innovazione, 28-29 Giugno 2007 Ferrara. Perugia: Morlacchi Editore: 303-315, 2007.

[54] Loucks, D. P., da Costa, J.R., *Decision Support Systems: Water Resources Planning,* NATO ASI Series, Sprinter Verlag, Berlin, 1991.

[55] Loucks, D.P., *Sustainable Water Resources Management,* Water International, vol.25, 3-10 March, 2000.

[56] Mays, L.W. Water Distribution Systems Handbook. New York: McGraw-Hill, 2000.

[57] McKenzie, R.S., Wegelin W. S. *Pressure Leakage Reduction: Public Private Partnership.* [online]. Proceedings of Leakage 2005, 12-14 September 2005 Halifax. Available from: http://waterloss2007.com/Leakage2005.com/unterseiten/final program.htm [Accessed 28 February 2009], 2005.

[58] Meier, R.W., Barkdoll, B.D. *Sampling design for network model calibration using genetic algorithms,* Jour. Water Res. Planning and Management, ASCE, 128(2), 130-139, July/August 2000, 2005.

[59] Milano, V., *Acquedotti,* Hoepli, Milano, 1996.

[60] Ministero dell'Ambiente e della Tutela del territorio, Comitato Comitato Italiano di lotta alla desertificazione, *Il deserto avanza:* 110 i Paesi a rischio, Roma, 2003.

[61] Ministero dell'Ambiente e della Tutela del territorio, Comitato per la Vigilanza sull'Uso delle Risorse Idriche, *Relazione annuale al Parlamento sullo stato dei servizi idrici* Anno 2004, Roma, 2005.

[62] Muraro, G. *La riforma dei servizi idrici in Italia,* Realt Nuova, 2, 2004.

[63] National Geographic Italia, *vol.10,* n.3, Milano, Settembre, 2002.

[64] OCSE. *Rapporto OCSE 2002 sulle politiche ambientali in Italia,* Parigi, 2002.

[65] OFWAT, ENVIRONMENT AGENCY, DEFRA, *Best Practice principles in the economic level of leakage calculation,* OFWAT Publications, 2002a.

[66] OFWAT, ENVIRONMENT AGENCY, DEFRA, *Leakage target setting for water companies in England and Wales,* OFWAT Publications, 2002b.

[67] Ostfeld, A., Shamir, U. Design reliable multiquality water-supply systems. Journal of Water Resources Planning and Management, 122 (5), 322-333, 1996.

[68] Ostfeld, A. *Water Distribution Systems Connectivity Analysis.* Journal of Water Resources Planning and Management, 131 (1), 58-66, 2005.

[69] Pezzinga, G., Gueli, R. *Optimal location of control valves in pipe networks by genetic algorithm*, Journal of Water Resources Planning and Management, ASCE, January/February pg. 65-67, 1999.

[70] Pezzinga, G. *Modelli per la regolazione della piezometrica ai fini della riduzione delle perdite in rete*, Atti del XXIV Convegno di Idraulica e Costruzioni Idrauliche, Napoli, Vol. II, T3b25-T3b38, 1994.

[71] Pezzinga, G., Pititto, G. *Combined optimization of pipes and control valves in water distribution network*, Journal of Hydraulic Research, vol. 43 n6, pag. 667-676, IAHR, 2005.

[72] Press, W.H., Teukolsky, S.A., Vetterling, W.T., Flannery, B.P., *Numerical Recipes in C: the art of scientific computing*, Second Edition, Cambridge University Press, 1992.

[73] Ravenna C., Favilli, R., Braccini G., *L'informazione Geografica fattore strategico nell'evoluzione delle Public Utilities*, MondoGIS, 34:25-29, 2003.

[74] Roger D. *Reducing leakage in Jakarta, Indonesia*, [online]. Proceedings of Leakage 2005, 12-14 September 2005 Halifax. Available from: http://waterloss2007.com/Leakage2005.com/unterseiten/final program.htm [Accessed 28 February 2009], 2005.

[75] Rossman, L.A. *EPANET2 Users Manual*. Cincinnati (OH): US E.P.A, 2000.

[76] Rudolph, G. *Convergence analysis of Canonical Genetic Algorithms*, IEEE Transactions on Neural Networks, 5(1), 96-101, 1994.

[77] Savic, D.A., G.A. *Walters. An Evolution Program for Optimal Pressure Regulation in Water Distribution Networks*, Engineering Optimization, 24 (3), 197-219, 1995.

[78] Sendil U., Al-Dhowalia, K. H. *Relationship between pressure and leakage in a water distribution network*, Proc. of the AWWA Annual Conference, Vancouver (B.C.), 665-6 85,1992.

[79] Shamir, U., Howard, C.D. *Water Distribution Systems Analysis*, Journal of Hydraulic Division, 94 (1), 219-234, 1968.

[80] Shun E., Park H., Park C., Hyun I. *A Case Study of Leakage Management in the City of Busan, Korea* [online]. Proceedings of Leakage 2005, 12-14 September 2005 Halifax. Available from: http://waterloss2007.com/Leakage2005.com/unterseiten/final program.htm [Accessed 28 February 2009], 2005.

[81] Siek, J.G., Lee, L-Q., Lumsdaine, A., *The Boost Graph Library User Guide and Reference Manual*. Upper Saddle River (NJ): Pearson Education, 2002.

[82] Skiena, S. *Implementing Discrete Mathematics: Combinatorics and Graph Theory with Mathematica*. Redwood City (CA): Addison-Wesley, 1990.

[83] Stanic, M., Avakumovic, D., Kapelan, Z. *Evolutionary algorithm for determining optimal layout of water distribution networks*, In: V. Babovic and L.C. Larsen, eds. Third International Conference on Hydroinformatics, 24-26 August 1998 Copenhagen. Lisse: A.A. Balkema, 901-908, 1998.

[84] Sturm, R. and Thornton, J. *Proactive Leakage Management using District Meter Areas (DMA) and Pressure Management - Is it applicable in North America?* [online]. Proceedings of Leakage 2005, 12-14 September 2005 Halifax. Available from: http://waterloss2007.com/Leakage2005.com/unterseiten/final program.htm [Accessed 28 February 2009], 2005.

[85] Sugiyama, K., Tagawa, S. and Toda, M. *Methods for Visual Understanding of Hierarchical Systems.* IEEE Transactions on Systems, Man and Cybernatics, SMC-11 (2), 109-125, 1981.

[86] Thornton, J. and Lambert, A. (2005). Progress in Practical Prediction of Pressure: Leakage, Pressure: Burst Frequency and Pressure: Consumption Relationships. Proc. of IWA Special Conference "Leakage 2005", Halifax, Nova Scotia, Canada

[87] Thornton, J., Shaw, M., Aguiar, M., Liemberger, R. *How Low Can You Go? A Practical Approach to Pressure Control in Low Pressure Systems,* [online]. Proceedings of Leakage 2005, 12-14 September 2005 Halifax. Available from: http://waterloss2007.com/Leakage2005.com/unterseiten/final program.htm [Accessed 28 February 2009], 2005.

[88] Todini, E., Pilati S. *A gradient algorithm for the analysis of pipe networks.* In: B. Coulbeck and O. Choun-Hou, eds. Computer Application in Water Supply, Vol I - System Analysis and Simulation. London: John Wiley & Sons, 1-20, 1988.

[89] Todini, E. *Looped water distribution networks design using a resilience index based heuristic approach,* Urban Water, 2, 115-122, 2000.

[90] Tzatchkov, V.G. Alcocer-Yamanaka, V.H. Ortiz, V.B. *Graph theory based algorithms for water distribution network sectorization projects.* In: S.G. Buchberger, R.M. Clark, W.M., 2006. Grayman and J.G. Uber, eds, 8th Annual Water Distribution Systems Analysis Symposium, August 27-30 2006 Cincinnati (USA). Reston (VA): ASCE, 323-330, 2006.

[91] UNESCO Working Group M.IV. *Sustainability Criteria for Water Resource Systems*, Cambridge University Press, 1999.

[92] UNESCO, Working Group M.IV, *Sustainability and optimization models.* Journal of Water Resources and Planning Management, 119(4): 455-472, 1993.

[93] United Nation, *Dublin Statemant*, Dublino, 2002a.

[94] United Nation, *Report of the Water Summit on Sustainable Development.* Johannesburg, South Africa, 26 August-4 Septempber. United Nations, New York, 2002b.

[95] Vitkovsky, J.P., Ligget, J.A., Simpson, A.R., Lambert, M., *Optimal Measurement Site Locations for Inverse Transient Analysis in Pipe Networks,* Journal of water resources planning and menagement, ASCE, Nov/Dic 2003.

[96] Wagner J.M., Shamir, U., Marks, D.H. *Water distribution reliability: analytical methods.* Journal of Water Resources Planning and Management, 114 (3), 253-274, 1988a.

[97] Wagner, J.M., Shamir, U., Marks, D.H. *Water distribution reliability: simulation methods.* Journal of Water Resources Planning and Management, 114 (3), 276-293, 1988b.

[98] Walters, G.A. and Savic, D.A., *Recent Applications of Genetic Algorithms to Water System Design*, Blain, W.R. (ed.), Hydraulic Engineering Software VI. Computational Mechanics Publications, Southampton, UK, 143-152, 1996.

[99] Warfield, J.Crossing Theory and Hierarchy Mapping. *IEEE Transactions on Systems, Man and Cybernatics,* SMC-7 (7), 502-523, 1977.

[100] Water Authorities Association and Water Research Centre. *Leakage Control Policy and Practice. Technical Working Group on Waste of Water.* London: WRc Group, 1985.

[101] UK Water Industry Research Ltd. *A Manual of DMA Practice,* London: UK Water Industry Research, 1999.

[102] World Water Council. *Third World Water Forum*, Kyoto, Japan, March, 2003.

[103] WRc plc, *Metodologia WRc di controllo permanente delle perdite in rete*, Wrc, LaboratoRI s.p.a., 2002.

[104] Wrc/WSA/WCA Engineering and Operations Committee. *Managing Leakage: UK Water Industry Managing Leakage*, Report A-J. London: WRc/WSA/WCA,1994.

[105] Wu, Z.Y., Simpson, A. R., *A self-adaptive boundary search genetic algorithm and its application to water distribution systems,* Journal Of Hydraulic Research, Vol. 40, No. 2, 2002.

GPSR Compliance

The European Union's (EU) General Product Safety Regulation (GPSR) is a set of rules that requires consumer products to be safe and our obligations to ensure this.

If you have any concerns about our products, you can contact us on ProductSafety@springernature.com

In case Publisher is established outside the EU, the EU authorized representative is:

Springer Nature Customer Service Center GmbH
Europaplatz 3
69115 Heidelberg, Germany

Batch number: 10726521

Printed by Printforce, the Netherlands